Challenges to Astronomy and Astrophysics:

Working Documents of the Astronomy Survey Committee

Working Groups

Astronomy Survey Committee

Commission on Physical Sciences,
Mathematics, and Resources

National Research Council

NATIONAL ACADEMY PRESS
Washington, D.C.   1983

Front Cover Map of radio emission from the galaxy 3C449 recorded by the Very Large Array (VLA) radio telescope of the National Radio Astronomy Observatory near Socorro, New Mexico. The map reveals highly collimated jets of matter connecting an unresolved galactic nucleus to outlying "lobes" of ejected gas. (Photo courtesy of the National Radio Astronomy Observatory)

Library of Congress Catalog Card Number 83-60509
International Standard Book Number 0-309-03335-7

Available from

NATIONAL ACADEMY PRESS
2101 Constitution Avenue, N.W.
Washington, D.C. 20418

Printed in the United States of America

Commission on Physical Sciences, Mathematics, and Resources

HERBERT FRIEDMAN, National Research Council, <u>Cochairman</u>
ROBERT M. WHITE, University Corporation for Atmospheric
    Research, <u>Cochairman</u>
STANLEY I. AUERBACH, Oak Ridge National Laboratory
ELKAN R. BLOUT, Harvard Medical School
WILLIAM BROWDER, Princeton University
BERNARD F. BURKE, Massachusetts Institute of Technology
HERMAN CHERNOFF, Massachusetts Institute of Technology
WALTER R. ECKELMANN, Exxon Corporation
JOSEPH L. FISHER, Office of the Governor, Commonwealth of
    Virginia
JAMES C. FLETCHER, University of Pittsburgh
WILLIAM A. FOWLER, California Institute of Technology
GERHART FRIEDLANDER, Brookhaven National Laboratory
EDWARD A. FRIEMAN, Science Applications, Inc.
EDWARD D. GOLDBERG, Scripps Institution of Oceanography
KONRAD B. KRAUSKOPF, Stanford University
CHARLES J. MANKIN, Oklahoma Geological Survey
WALTER H. MUNK, University of California, San Diego
NORTON NELSON, New York University Medical Center
DANIEL A. OKUN, University of North Carolina
GEORGE E. PAKE, Xerox Research Center
CHARLES K. REED, National Research Council
HATTEN S. YODER, JR., Carnegie Institution of Washington

RAPHAEL G. KASPER, <u>Executive Director</u>

Astronomy Survey Committee

GEORGE B. FIELD, Harvard-Smithsonian Center for Astrophysics, Chairman
MICHAEL J.S. BELTON, Kitt Peak National Observatory
E. MARGARET BURBIDGE, University of California, San Diego
GEORGE W. CLARK, Massachusetts Institute of Technology
S. M. FABER, University of California, Santa Cruz
CARL E. FICHTEL, NASA Goddard Space Flight Center
ROBERT D. GEHRZ, University of Wyoming
EDWARD J. GROTH, Princeton University
JAMES E. GUNN, Princeton University
DAVID HEESCHEN, National Radio Astronomy Observatory
RICHARD C. HENRY, The Johns Hopkins University
RICHARD A. McCRAY, Joint Institute for Laboratory Astrophysics and the University of Colorado
JEREMIAH OSTRIKER, Princeton University
EUGENE N. PARKER, University of Chicago
MAARTEN SCHMIDT, California Institute of Technology
HARLAN J. SMITH, University of Texas, Austin
STEPHEN E. STROM, Kitt Peak National Observatory (ex officio)
PATRICK THADDEUS, NASA Goddard Institute for Space Studies and Columbia University
CHARLES H. TOWNES, University California, Berkeley
ARTHUR B.C. WALKER, Stanford University
E. JOSEPH WAMPLER, University of California, Santa Cruz

PAUL BLANCHARD, Executive Secretary
DALE Z. RINKEL, Administrative Secretary

WORKING GROUP ON SOLAR PHYSICS

Arthur B.C. Walker, Stanford University, <u>Chairman</u>
John W. Harvey, Kitt Peak National Observatory
Thomas E. Holzer, National Center for Atmospheric Research
Jeffrey L. Linsky, Joint Institute for Laboratory Astrophysics
    and the University of Colorado
Eugene N. Parker, University of Chicago
Roger K. Ulrich, University of California, Los Angeles
Gerard Van Hoven, University of California, Irvine
George L. Withbroe, Harvard-Smithsonian Center for Astrophysics

Consultants

Hugh S. Hudson, University of California, San Diego
Stuart D. Jordan, NASA Goddard Space Flight Center
Mukul R. Kundu, University of Maryland
Jack B. Zirker, Sacramento Peak Observatory

WORKING GROUP ON PLANETARY SCIENCE

Michael J.S. Belton, Kitt Peak National Observatory, <u>Chairman</u>
John J. Caldwell, State University of New York, Stony Brook
Donald M. Hunten, University of Arizona
Torrence V. Johnson, Jet Propulsion Laboratory
David Morrison, University of Hawaii
Tobias C. Owen, State University at New York, Stony Brook
Stanton J. Peale, University of California, Santa Barbara
Gordon H. Pettengill, Massachusetts Institute of Technology
James B. Pollack, NASA Ames Research Center

WORKING GROUP ON GALACTIC ASTRONOMY

Robert D. Gehrz, University of Wyoming, <u>Chairman</u>
David Black, NASA Ames Research Center
W. Butler Burton, University of Minnesota
Duane F. Carbon, Kitt Peak National Observatory
Judith G. Cohen, California Institute of Technology
Pierre Demarque, Yale University
Frederick K. Lamb, University of Illinois, Urbana
Bruce Margon, University of Washington, Seattle
Philip Solomon, State University of New York, Stony Brook
Sidney van den Bergh, Dominion Astrophysical Observatory
Peter O. Vandervoort, University of Chicago

Consultants

Richard A. McCray, Joint Institute for Laboratory Astrophysics
    and the University of Colorado
Christopher F. McKee, University of California, Berkeley
Leonard Searle, Carnegie Institution of Washington

WORKING GROUP ON EXTRAGALACTIC ASTRONOMY

S. M. Faber, University of California, Santa Cruz, <u>Chairman</u>
Christopher McKee, University of California, Berkeley
Frazer Owen, National Radio Astronomy Observatory
P. James E. Peebles, Princeton University
Joseph Silk, University of California, Berkeley
Harvey Tananbaum, Harvard-Smithsonian Center for Astrophysics
Alar Toomre, Massachusetts Institute of Technology
James W. Truran, University of Illinois, Urbana
Ray J. Weymann, University of Arizona

James E. Gunn, Princeton University, <u>ex officio</u>
Jeremiah Ostriker, Princeton University, <u>ex officio</u>

Consultant

Beatrice M. Tinsley, Yale University

WORKING GROUP ON RELATED AREAS OF SCIENCE

James E. Gunn, Princeton University, <u>Chairman</u>
Douglas Eardley, Harvard-Smithsonian Center for Astrophysics
Peter Gilman, National Center for Atmospheric Research
Russell M. Kulsrud, Princeton University
David Pines, University of Illinois, Urbana
Gerald J. Wasserburg, California Institute of Technology
William D. Watson, University of Illinois, Urbana
Steven Weinberg, Harvard University
Stan E. Woosley, University of California, Santa Cruz

WORKING GROUP ON ASTROMETRY

Gart Westerhout, U.S. Naval Observatory, <u>Chairman</u>
Heinrich K. Eichhorn, University of Florida, Gainesville
George D. Gatewood, Allegheny Observatory
James Hughes, U.S. Naval Observatory
William H. Jefferys, University of Texas, Austin
Ivan R. King, University of California, Berkeley
William F. Van Altena, Yale University

WORKING GROUP ON THE SEARCH FOR EXTRATERRESTRIAL INTELLIGENCE

Harlan J. Smith, University of Texas, Austin, <u>Chairman</u>
Frank Drake, Cornell University
James E. Gunn, Princeton University
David Heeschen, National Radio Astronomy Observatory
Noel W. Hinners, Smithsonian Institution
Jeremiah Ostriker, Princeton University
Patrick Thaddeus, NASA Goddard Institute for Space Studies
    and Columbia University
Charles H. Townes, University of California, Berkeley
Benjamin M. Zuckerman, University of Maryland

Consultants

George D. Gatewood, Allegheny Observatory
Michael Hart, Trinity University, San Antonio
Michael D. Papagiannis, Boston University

Preface

This volume contains the documents submitted by seven Working
Groups to the Astronomy Survey Committee, whose reports,
Astronomy and Astrophysics for the 1980's, Volume 1: Report of
the Astronomy Survey Committee (1982) and Volume 2: Reports of
the Panels (1983), are being published contemporaneously
(National Academy Press, Washington, D.C.). Chapter 3 of the
Committee's report, "Frontiers of Astrophysics," draws heavily
on the material in the present volume.

The National Academy of Sciences charged the Astronomy
Survey Committee with assessing the opportunities for progress
across the entire range of astronomical research and with
making recommendations concerning the programs and facilities
needed to meet those opportunities. In order to carry out the
first part of its charge, the Committee asked a number of
experts in each area of astronomy and astrophysics to form
Working Groups to consider the state of knowledge in each area
and to identify the most important scientific questions in
these areas for the 1980's.

The reports of the Working Groups provided key inputs not
only to the Committee, but also to the various technique-
oriented Panels, which the Committee established to propose
recommendations for programs and facilities. The Working
Groups' reports proved to be of such a high quality that the
Committee decided to sponsor their publication as an
independent supplement to the Committee report.

The activities of the Working Groups and Panels were
mutually supportive. The fundamental scientific questions
identified by the Working Groups, which require a variety of
techniques to address, were helpful to the Panels in framing

programs in their areas of technical expertise that would provide maximal scientific return in a variety of different areas. Similarly, the Panel reports were reviewed by the Working Groups to ensure that the Panel recommendations were responsive to the scientific opportunities identified in the Working Group reports.

In the case of observational astronomy, the Committee limited its recommendations to programs designed to obtain information about astronomical objects by remote sensing from the Earth or its vicinity. The Committee thus specifically excluded from consideration recommendations for instruments on spacecraft designed to escape Earth orbit, for programs to study the Earth itself, for study of samples of matter originating beyond the Earth, and for instruments to test the predictions of different theories of gravitation.

To varying degrees, the Working Groups nevertheless found it essential to consider some of these issues in order to provide a complete picture of the opportunities they perceived. Such discussions furnished helpful background for the Committee; however, they should not be regarded as endorsements by the Committee of programs explicitly excluded from consideration for recommendations by the Committee.

The Working Group on Solar Physics provided a comprehensive discussion of the opportunities facing solar research and went beyond that in proposing a coherent strategy for solar research utilizing both ground- and space-based techniques at a wide variety of wavelengths. Formulation of such a strategy, made possible because the solar-physics community is well organized for undertaking coordinated research projects related to its single object of study, resulted in recommendations for solar research that were examined by the experts in each of the technique-oriented Panels.

The Working Group on Planetary Science, informed of the Committee's decision not to consider recommendations for instrumentation aboard spacecraft designed to escape Earth orbit, left this task to other duly constituted committees. However, its report does consider in depth many other ways of acquiring information about the planets, including telescopic observations from the ground and from Earth orbit, radar studies, analysis of lunar and meteoritic samples, laboratory spectroscopy and materials studies, and computer modeling and theoretical studies. A helpful survey of proposed deep-space planetary missions was also provided.

Galactic and extragalactic astronomy includes studies of a wide variety of objects, ranging from interstellar grains to galaxies, and from red dwarf stars to quasars. In spite of the fact that virtually all astronomical techniques are used in such studies, the corresponding Working Groups were successful in identifying broad themes that will permeate astronomical research in the 1980's. These discussions provided the scien-

tific background needed by the Committee in considering the
assignment of priorities to projects proposed by the Panels.

The Committee found that three research areas required
special attention if the opportunities in them were not to be
overlooked. A Working Group on Related Areas of Science
surveyed developments in other fields of importance to
astronomy and brought to the Committee's attention possible
future directions. A Working Group on Astrometry discussed the
development of this field, which is such an important underpin-
ning to all other work in astronomy. Finally, a Working Group
on the Search for Extraterrestrial Intelligence was constituted
largely (but not wholly) from the membership of the Committee
itself, in order to deal with the issues facing this somewhat
unconventional field.

Each of the Working Groups produced a thoughtful and timely
report, for which the Committee is grateful and of which they
and the scientific community can be proud.

George B. Field, Chairman
Astronomy Survey Committee

Contents

Challenges to
Astronomy and
Astrophysics

1

Solar Physics

I. INTRODUCTION

The past few years have seen an increasingly close coupling
between solar and nonsolar astronomy, brought about largely by
our increasing ability to study Galactic and extragalactic phenomena in greater detail and with greater sensitivity over the
entire electromagnetic spectrum. Such phenomena as hydromagnetic flows (winds), coronas, stellar magnetism, activity
cycles, and particle acceleration--which until recently could
be readily observed only in the Sun and in a few Galactic and
extragalactic objects in which extreme conditions of temperature or gravitational or magnetic-field strength prevailed--can
now be studied in a wide range of stars and galaxies. The
Working Group on Solar Physics believes that both solar physics
and the other disciplines of astronomy will benefit from this
more intimate interaction, as we point out later in this
chapter.

However, solar physics differs in two major respects from
the other disciplines of astronomy. First, solar physics has
achieved the complete integration of its scientific program
over the full range of observational techniques available to
astronomers. The resulting observations now cover the entire
electromagnetic spectrum, from flare-excited nuclear gamma-ray
lines at millions of electron volts in energy to low-frequency
radio waves (at decametric and kilometric wavelengths) generated by the propagation of electrons in the corona and the interplanetary medium; the spectrum of neutrinos from thermonuclear
reactions in the solar core; and in situ measurements of low-
energy plasma, high-energy cosmic rays, and of electric and

1

magnetic fields in the interplanetary medium and the corona. Because of the intimate connection of these observations to each other, and the need in most cases for the simultaneity of these observations, the Working Group supports an integrated theoretical and observational strategy for solar physics during the 1980's that may be pursued through the closely coordinated set of programmatic opportunities outlined toward the end of this chapter.

The second way in which solar physics differs from other disciplines of astronomy is its close connection to other areas of scientific inquiry, particularly to heliospheric physics, space physics, and the study of the Earth's climate. The Working Group therefore felt it imperative to adopt a broad definition of solar physics, one that includes not only the study of the Sun itself but also the study of the heliosphere-- the entire volume of space (extending well beyond the solar system) that is dominated by the solar wind--together with the interaction of the solar radiative and particulate flux with planetary atmospheres, ionospheres, and magnetospheres. This broad definition has important implications for the relevance to solar physics of interplanetary space missions intended for in situ studies of the heliosphere. Such missions are therefore important components of the programmatic opportunities recognized by the Working Group.

Finally, the Working Group considered a number of issues relating to the institutional arrangements for solar-physics research and education that have arisen over the past two decades, and we have included a brief discussion of them at the end of this chapter. These issues are important, but they are also complex; their resolution will require a concerted response not only by the relevant funding agencies--the National Science Foundation in (NSF), the National Aeronautics and Space Administration (NASA), and the Department of Defense (DOD)--but also by the community itself, as represented by the Solar Physics Division of the American Astronomical Society, in the very near future.

II. OVERVIEW AND GENERAL CONCLUSIONS

A. Definitions and Major Themes

Solar physics has evolved so rapidly in the past decade as to require additional clarification of the aim and intent of this chapter. Several connotations of the term "solar physics" are currently in vogue. Some scientists accord it only the narrowest and most literal meaning, i.e., the study of a particular object that happens to be the closest star; others, however, use the term "solar physics" more broadly to include the varied phenomena that occur within the interplanetary medium, together

with the interaction of the solar wind with planetary magnetospheres and atmospheres. They go on to point out that the Sun
provides a laboratory for the study of basic evolutionary
processes occurring in stellar interiors, of cyclic and
transient phenomena in stellar atmospheres, and of violent
phenomena involving the acceleration of particles to very high
energies that take place in supernovae and in active galaxies--
all at the level of basic plasma processes that no other readily
observable situation provides. The Sun, a star of average size
and luminosity, is our most accessible experimental window into
the vast and varied activities of other stars and galaxies too
distant to be resolved. The recent report by the Space Science
Board's Committee on Solar and Space Physics (Solar-System Space
Physics in the 1980's: A Research Strategy, National Academy of
Sciences, Washington, D.C., 1980) gives an excellent and
extensive presentation of this view.

The present Working Group on Solar Physics favors a variation of this latter attitude; we see solar physics as a fundamental inquiry into the physics of the remarkable large-scale
behavior of ionized gases in gravitational and electromagnetic
fields. This inquiry has three major themes:

1. The development of observational techniques that can
directly probe the generation and transport of energy and the
composition and motion of matter in a stellar interior, therefore providing direct experimental tests of models of stellar
structure and evolution.

2. The detailed study of the various active phenomena
that can be observed in the solar atmosphere and that have only
recently become observable in other stars. These active
phenomena are consequences of the generation of magnetic fields
and of large-scale circulation, which are themselves by-products
of energy-transport processes operating in the convective
envelope of the Sun. Each active phenomenon--such as a flare,
the solar wind, or the sunspot cycle--is in fact not a single
effect but rather a combination of effects all operating
together. This cooperative aspect of solar activity cannot be
duplicated in the terrestrial laboratory because many of the
individual effects do not function on so small a scale.

3. The study of the three-dimensional structure and
dynamics of the corona and interplanetary medium (together
frequently referred to as the heliosphere) and their interaction with planetary atmospheres and the interstellar medium.
The realization that changes in the structure of the interplanetary medium (brought about by changes in the level of
solar activity) are closely correlated with major changes in
climatic conditions on the Earth suggests that heliospheric
physics may become a significant factor in issues that are of
great practical importance, as well as of intrinsic scientific
interest.

Progress has been steady and profoundly rewarding to those who have pursued these inquiries. Some of the more remarkable discoveries and achievements of the past decade include:

• The direct study of thermonuclear processes occurring in the solar core through observations of neutrinos. The disagreement of the observed flux with that predicted by standard models has resulted in the re-examination of the models and the planning of new observations that should provide more definitive interpretations.

• The discovery that the 5-min oscillations of the solar atmosphere are a global phenomenon that can be used as a probe of the structure and dynamical behavior of the solar interior.

• The discovery that the damping of solar atmospheric waves driven by convection cannot account for the energy required to heat the corona and drive the solar wind. The observation of coronal phenomena in main-sequence stars in every part of the H-R diagram has reinforced the conclusion, drawn from solar observations, that magnetic effects underlie active phenomena in stellar atmospheres.

• The confirmation of the evidence (provided initially by seventeenth-century observations) that the sunspot cycle and associated active phenomena were largely absent for a period of 70 years in the seventeenth century. This episode is known as the Maunder Minimum. We now know that such interruptions, along with periods of heightened activity, occur quasi-periodically and that there is a strong correlation between these periods of inactivity or hyperactivity and the occurrence of climatic changes on the Earth.

• The recognition that the energy released during the impulsive phase of a solar flare is largely or entirely contained in nonthermal particles accelerated during magnetic-reconnection processes in the coronal field.

• The demonstration that the large-scale solar magnetic field is organized into two distinct types of structures: magnetically closed regions, in which hot plasma confined in loops largely generates the x-ray corona; and magnetically open regions, the so-called "coronal holes," which are the source of high-speed streams in the solar wind.

• The discovery that, when viewed on a fine scale, the solar magnetic field is subdivided into individual flux tubes with field strengths exceeding 1000 gauss and with a physical size smaller than can be resolved by any present telescope.

It is clear that these discoveries are merely the beginning of a more detailed physical understanding of stellar structure and evolution and of the various active phenomena that occur in stellar atmospheres, and they also provide the theoretical basis for understanding the heliosphere and its interaction with the Earth and other planets; obviously, much nevertheless remains to be discovered and explained.

B. Summary of Major Scientific Problems and Programmatic Initiatives for the 1980's

Guided by this expanded definition of the meaning and scope of solar physics, the Working Group identified six scientific problems that address major issues related to the three themes introduced above. We believe that the pursuit of these problems can be especially productive during the 1980's because of the advent of more powerful computational capabilities and the potential for the development of observational facilities in space with greatly improved spatial resolution and sensitivity, represented by the Space Shuttle.

These problems can be posed as a series of questions. We present these major questions below and discuss their implications, in both scientific and programmatic terms, in the remainder of this chapter.

1. What are the fundamental properties of the solar core? In particular, what is its rotation rate, chemical composition, and temperature distribution, and what is the detailed process of nuclear-energy generation? How do these properties relate to current theories of stellar evolution?

2. What is the hydrodynamic structure of the solar convection zone (particularly the character of the solar dynamo), the nature of large-scale circulation, and the implications of very-long-period global oscillations?

3. What physical mechanisms drive the solar activity cycle, what resulting variations in the solar radiative and particulate output follow on various time scales, and what is the effect of this variability on the Earth's upper atmosphere? How do these relate to activity and variability on other stars?

4. What processes, involving small-scale velocity and magnetic fields and various wave modes, determine the thermodynamic structure and dynamics of the solar photosphere, chromosphere, and corona, and what are the implications of such processes for stellar atmospheres in general?

5. What are the basic plasma-physics processes responsible for metastable energy storage, magnetic reconnection, particle acceleration, and energy deposition in solar flares and related nonthermal phenomena? What are the implications of these for other high-energy processes in the Universe?

6. What are the large-scale structure and plasma dynamics of the solar corona, including the processes involved in heating various coronal structures and initiating the solar wind? What is the origin of coronal transients? What is the three-dimensional structure of the interplanetary medium, and what are its implications for cosmic-ray modulation and for the modulation of planetary atmospheres and ionospheres? What are the implications for stellar coronas and winds and other astrophysical flows?

While the questions above are posed primarily in terms of solar phenomena, we believe that continued progress of all branches of astrophysics is best served by an integrated approach to understanding these basic physical processes through the comparative study of the Sun and other astrophysical objects. Such comparative studies benefit both solar physics and the other major disciplines of astrophysics, the insights gained in one subdiscipline having important consequences for many other subdisciplines. In the next section we discuss the close connection between the physical processes that are the main focus of current solar research and the other astrophysical settings in which these same processes must operate.

Section IV presents a comprehensive scientific program designed to address the six major scientific problems outlined earlier and includes a discussion of the improved theoretical and observational capabilities that will be required to carry out this program. The principal new initiatives identified by the Solar Physics Working Group as having high priority for solar physics during the 1980's are the following:

1. The deployment of an Advanced Solar Observatory (ASO)--containing an ensemble of instruments capable of simultaneous, high-resolution observations of the solar atmosphere over the full spectral range of wavelengths from gamma rays to the infrared spectral region--upon a Shuttle-serviced space platform. The development of ASO should follow an evolutionary path that incorporates major instruments developed previously for flight aboard the Space Shuttle, such as the Solar Optical Telescope (SOT) recommended by the Space Science Board's Committee on Solar and Space Physics. The Working Group regards the development of ASO as deserving of the highest priority among solar programs for the 1980's.

2. A comprehensive program for the study of the solar convection zone and of solar activity. Such a program should include the study of the dynamical behavior of the convection zone initially from the ground and subsequently from a Solar Interior Dynamics Mission (SIDM) in space, together with the study of the dynamical behavior and evolution of the corona from a dedicated Explorer mission, the Solar Coronal Explorer (SCE).

3. The launch of an entirely new kind of solar mission, the Star Probe, for a close encounter with the Sun, in order to achieve the following scientific objectives:

a. The direct, in situ study of the region in which the solar wind is accelerated;

b. Measurements of the internal mass distribution and rotation profile of the Sun;

c. Study of the structure of the solar atmosphere and associated dynamical processes at the extremely high spatial resolution (7-10 km on the solar surface); and

d. Study of the large-scale structure of the corona and the heliosphere through both in situ and remote observations.

4. The expansion of continuing programs to measure the flux of neutrinos from the solar core, particularly through the implementation of an experiment employing large quantities of ^{71}Ga for an accurate determination of the low-energy solar-neutrino flux.

5. A comprehensive program for the study of active phenomena in other stars through both ground-based and spacecraft observations, with strong emphasis on comparative solar/stellar observations.

6. An increased emphasis of the role of theory and modeling in solar physics.

Section V presents a more detailed account of the scientific issues to be addressed during the coming decade, whereas Section VI furnishes a more complete description and justification of the programs outlined above. A final section discusses institutional issues of concern in the coming decade.

However, before proceeding to these topics, we illustrate the close connection between solar physics and other branches of astrophysics through a more detailed discussion of one of our major themes: active phenomena in stellar atmospheres.

C. Example: Solar and Stellar Activity

Four of the major problems enumerated above (2, 3, 4, and 5) involve different aspects of the study of active phenomena in stellar atmospheres, a set of inquiries that we might call the physics of stellar activity or stellar plasma dynamics, with applications to the Sun and other stars. These terms take on added meaning in connection with our rapidly expanding ability to observe activity in other main-sequence stars, adding a new dimension to the physics of stellar activity, which until recently had been concerned largely with phenomena (such as stellar pulsations) that occur only during brief and atypical evolutionary phases. Indeed, a particularly fascinating feature of stars is their remarkable repertoire of activity, which displays variations throughout the long main-sequence lifetime and in the subsequent giant and dwarf stages. It is these manifestations of activity that have maintained astronomical interest in the Sun; and it is the activity of other stars that has been such a revelation in recent years, showing that magnetic fields, chromospheric activity, flare-like effects, coronal x rays, and stellar winds are part of the normal daily life of essentially all classes of stars. It is a simple fact that the atmospheres of all stars are in states that are far removed from both local thermodynamic equilibrium (LTE) and hydrostatic equilibrium.

How, then, do we go about understanding the complex activity of the stars? So far as we are aware, the basic equations of physics—those of Newton, Maxwell, and Dirac—give a complete and accurate description of the behavior of atoms and fields in stellar atmospheres. On the other hand, we must ourselves construct the mathematical solutions to these equations. In the presence of a large number of atoms, the variety of possible solutions is staggering; we have had little or no luck in anticipating a priori the various classes of solutions that are important in real stars.

We must therefore rely to a large extent on an empirical approach, first observing each phenomenon of interest in sufficient detail to recognize the basic physical processes that underlie this phenomenon. Only then can we construct a solution to the basic equations of physics that can explain such phenomena in terms of the fundamental laws of physics. It is in this step of the process that solar observations play so basic a role, for in many instances only the Sun allows observations with sufficient resolution and sensitivity to permit identification of the basic physical processes that play a role.

It should be emphasized that, largely as a result of the study of solar phenomena, there is now a firm theoretical understanding of many of the component effects that together make up stellar activity. For example, several special cases of neutral-point rapid magnetic reconnection have now been solved, so that we have a basis for judging the range of reconnection rates that are possible. The turbulent diffusion of magnetic fields (including the newly discovered negative turbulent diffusion) has been formally investigated in several limiting cases. The theory of generation of magnetic fields in a wide variety of fluid motions, ranging from small-scale turbulence to stationary flows, has been put on a formal basis, and the theory of convection and circulation in a rotating stratified spherical body has seen significant progress. A formal theory of coronal expansion and stellar winds, with coronal temperature as a parameter, has been applied to a variety of special circumstances to show the wide range of possible effects. We can be particularly proud of the development of the non-LTE theory of radiative transfer in stellar atmospheres, which is now available for interpreting the spectra of other stars and of nebulae and quasars; applications of this theory to a number of stars has already been remarkably successful. There are now firm theoretical models for a variety of particle-acceleration mechanisms suggested by studies of solar flares; these are available for application to other astrophysical sources of fast particles.

The diverse activity currently visible on the Sun and on more distant stars collectively exhibit the extraordinary range of effects that must be explained. Other classes of stars complement the solar laboratory by exhibiting more extreme activity over a wide variety of conditions. The spectrum of activity in the other stars shows the latitude of variation that may be expected in the Sun throughout its slow evolution. Long-term variations in the level of solar activity (and presumably in the activity of other stars) have become apparent from studies of the average strength of the solar wind over the last 10^9 years, as revealed by analysis of rocks exposed on the surface of the Moon and from recent studies of ^{14}C production.

In summary, the physics of stellar activity is pushing forward on an expanded front, with detailed and precise studies of the activity on the Sun in the vanguard, flanked by studies of other stars of all types.

D. Remarks on Continuing and Related Programs

The Solar Physics Working Group wishes to emphasize strongly the importance of continuing and related programs--including those planned or under development but not yet completed--to the scientific program developed here.

In each of the major theme areas that we have defined, the vigorous operation of continuing and level-of-effort programs is absolutely essential and has been taken by the Working Group as the foundation upon which the solar-physics program of the coming decade is to be constructed. Of particular importance to solar physics are the NSF grants program; NSF support to programs at the National Astronomy Centers; and the Research and Analysis, Suborbital, Spacelab, and Explorer programs of NASA at existing or augmented levels.

Furthermore, the currently approved major programs of NASA are regarded by the Working Group as the framework for the necessary scientific and technical advances needed for the effective development and utilization of the major new initiatives that we have identified. In this connection, the Working Group wishes to emphasize the importance of adequate funding for analysis of the results of the Solar Maximum Mission (SMM); of the timely development of major Shuttle-science facilities through the Spacelab program, particularly the SOT; and of the International Solar Polar Mission (ISPM). (Following completion of the present report, NASA announced suspension of plans to provide a U.S. spacecraft as part of ISPM. However, the agency expressed continued support for the European ISPM spacecraft and subsequently initiated study of a redefined U.S. spacecraft as a possibility for inclusion in an augmented fiscal year 1983 NASA budget. The Working Group wishes strongly to reiterate its belief that a dual-spacecraft ISPM mission with imaging capability is essential to an orderly and effective program of solar and heliospheric studies.)

Finally, the Working Group wishes to point out the importance to solar physics of three interdisciplinary space missions
currently under study at NASA. These are the Origin of Plasmas
in the Earth's Neighborhood (OPEN) mission, reviewed and
endorsed by the Space Science Board's Committee on Solar and
Space Physics; the Solar Terrestrial Observatory (STO), a
Shuttle/space platform facility currently under preliminary
study at NASA; and the Advanced Interplanetary Explorer (AIP)
mission for the study of the transient-particle population of
the interplanetary medium.

III. THE NATURE OF SOLAR PHYSICS

As pointed out in the Introduction, solar physics is an open-
ended discipline with close connections to virtually every
branch of astrophysics and to several major subdisciplines of
physics and geophysics. Here we review these connections and
point out how the scientific objectives of solar physics and of
other disciplines complement each other.

A. The Sun as a Laboratory of Stellar Processes

1. Stellar Interiors and Nucleosynthesis

The theory of nucleosynthesis in stellar interiors has provided
the foundation for models of the dynamics of stellar interiors
and of stellar evolution. Direct experimental tests of these
general stellar models are at present beyond our reach, and
many crucial parameters involved in them (such as nuclear
reaction rates and opacities) are difficult to measure directly
in the laboratory.

However, direct tests of models of the solar interior can
be carried out by three techniques: the observation of the
energy spectrum and intensity of the solar-neutrino flux; the
observation of the solar mass distribution through measurements
of perturbations on the orbit of a close gravitational probe;
and the observation of the global oscillations of the Sun.
Although only the third of these techniques appears to be feasible for stars other than the Sun in the foreseeable future,
solar observations would appear to offer the principal opportunity to complement indirect tests of models of stellar interiors furnished by the study of stellar evolution with more
direct tests. Some of the major issues that these observations
can address are discussed below.

a. Interior Elemental Abundance Distribution

Current models of stellar evolution assume that stars initially have a uniform interior abundance and that the abun-

dance gradients established by thermonuclear reactions in the stellar core during the main-sequence lifetime are not strongly affected by mixing with material outside the core. However, an initial nonuniform composition--specifically, a core with lower abundances of elements heavier than helium--is not ruled out by present theories of star formation.

In the case of the Sun, an initial composition gradient could have developed if substantial accretion of matter occurred after the Sun developed a radiative core. Such accreted matter may well have undergone some degree of elemental segregation. Nor can significant mixing of the thermonuclear core with material outside the core be ruled out. The spectrum and flux of solar neutrinos, which can be interpreted to provide rates for the various thermonuclear reactions that occur in the Sun, provide a mechanism to test these hypothesis. Plans are now under way to supplement the measurement of the high-energy neutrino flux measured by the [37]Cl experiment with measurements by radiochemical detectors sensitive to low- and medium-energy neutrinos. Since the distribution of abundances in the solar interior can effect the internal temperature and mass equilibrium, the perturbations of a close gravitation probe arising from a solar quadrupole moment would also place constraints on solar abundance models.

b. Solar Core Rotation

The theory of solar core rotation (and indeed of stellar core rotation) is sufficiently complex to permit considerable latitude in the possible rotation laws. While extremely rapid rotation rates (faster than 10 hours) have been ruled out on the basis of solar oblateness measurements, rotation rates as much as two or three times the surface rate are consistent with present observations. A rapidly rotating core below the convective envelope could provide a fundamental source of organized kinetic energy, which could well play a role in the solar dynamo. The presence of such a core could be detected by its effect on a close gravitational probe.

c. Dynamic and Static Structure at the Convective Envelope Base

Theory suggests that the largest convective elements or cells on the Sun should be most persistent. Such cells are expected to be associated with the inner boundary of the convection zone (except for a possible interior boundary layer where there may be a transition zone from moving matter to the quiescent state). There is a hint that these largest convective elements have been detected in the form of giant cells. They may also be at least partly responsible for persistent phenomena like coronal holes. An understanding of their size, lifetime, and velocity pattern would be extremely useful in understanding the dynamics of the solar convective envelope. Their depth and the depth of the inner boundary of the convective envelope will also be useful parameters. Ultimately, this question and the question of large-scale circulation are tied together, since the large-scale circulation may be the product of effects induced by the largest convection cells. There is now evidence for cells on other stars, specifically supergiants; however, these observations will remain at the limits of currently available resolving power even after Space Telescope becomes operational. Solar observations will continue to be at the center of efforts to understand convective processes in stars.

2. Energy Transport in Stars

The Sun is a unique object in astrophysics by virtue of its proximity and the diagnostic capability that that proximity allows. Furthermore, the solar radiative flux is essentially unattenuated by intervening matter (except for the terrestrial atmosphere), whereas interesting stellar emissions shortward of 912-$\overset{\circ}{A}$ wavelength and in the cores of ultraviolet resonance lines are heavily attenuated. As a consequence, many phenomena routinely studied on the Sun have not been observed in other stars, although they are presumably present. In particular, we are able to study in detail the radiative, convective, and magnetic processes that transport energy through the photosphere, and the tenuous chromosphere and corona. The basic physical mechanisms underlying these processes are of great interest in themselves, but they are of even greater interest to the astrophysicist, as these basic mechanisms must be ubiquitous throughout astrophysics and must therefore underlie a wide range of observed phenomena.

a. Radiation Transport and the Interpretation of Stellar Spectra

In the simplest approximation, astrophysical plasmas are presumed to be in local thermodynamic equilibrium (LTE), so that the degree of ionization depends only on density and temperature. Although the LTE presumption is commonly used in interpreting stellar spectra, it is known to be only rarely valid. In the solar context, various deviations from this ideal state have been studied in many physical situations. The helium ionization balance in the solar chromosphere is an example of photoionization/recombination equilibria that also may occur in T Tauri stars, x-ray binaries, Seyfert galaxies, and quasars. In the solar corona, ionization equilibria are often dominated by conditions that are probably valid also in other stellar coronas, in interstellar shock fronts, and in the x-ray emitting halos of clusters of galaxies.

More recent observations have made it necessary to develop a theory of nonlocal ionization balance in which the ionization equilibrium is modified by mass motions, thermal diffusion of ions, and penetration of electrons from elsewhere in the atmosphere. These theoretical models, which have already been observationally confirmed in the case of the solar wind, may help in understanding nonlocal and time-dependent equilibrium phenomena that occur in novae, supernovae, flare stars, shock fronts, and the hot component of the interstellar medium.

The Sun has also furnished physical insight into various mechanisms for spectral-line formation. A variety of types of excitation conditions that occur in stellar atmospheres have also been studied in solar contexts. For example, much of the study of spectral-line formation in the presence of systematic and random velocity fields and velocity gradients has derived its impetus from solar astrophysics. Solar research has also provided examples of radiative transfer in the presence of magnetic fields. These mechanisms are particularly relevant to extended gas and dust envelopes, Be stars, magnetic A stars, and Of stars.

In theoretical studies of heating mechanisms and energy balance in astrophysical plasmas, it is essential to understand radiative cooling. Detailed observations of transient solar phenomena provide a stringent test of the general cooling functions developed for optically thin astrophysical plasmas. At higher densities, some transitions become optically thick and the simple optically thin approximations are no longer valid. Radiative-loss calculations for such plasmas have been developed to understand the solar photosphere and chromosphere and are also utilized to study flare stars and stellar atmospheres in general.

b. Convection and Stellar Coronas and Winds

The granular structure of the Sun's photosphere has from the beginning been attributed to the process of convection--one of the dominant mechanisms of energy transport in stars. Analytical theories of convection have been developed, but only for simplified model atmospheres that do not yield the preferred length scales characteristic of the solar atmosphere. A detailed understanding of convection must be derived from the comparison of computer models of the solar atmosphere with observation.

Our understanding of the hydrodynamics of the extended solar atmosphere is much more detailed than that of any other astrophysical object. For example, the theory of the solar wind represents a major triumph of modern astrophysics, which has led in turn to the concepts of a "polar wind" from the Earth, of "stellar winds," and even of "galactic winds."

A central problem in contemporary solar research is the transport of energy between the spherically symmetric chromosphere (at 10^4 K) and the nonsymmetric corona (at temperatures in excess of 10^6 K). Very recently, coronas have been detected in an unexpectedly wide range of stars by x-ray observations from rockets and the ANS, HEAO-1, and Einstein x-ray observatory (HEAO-2) satellites. The Sun has been the testing ground for theories to account for the heating and physical properties of chromospheres and coronas. While it is generally believed that energy available in convective motions is converted into heat in the upper layers, the extent to which this heating is due to dissipation of waves generated by the convective motions or to the annihilation of magnetic fields produced by displacement of field lines embedded in the convection zone is at present unclear. Considerable effort is now under way to test both of these theoretical processes through comparisons with realistic models of solar chromospheric and coronal structures and the observed properties of the solar wind. Theoretical methods that successfully match solar observations can then be used to understand stellar data; and, conversely, the new stellar observations can further refine our theoretical methods and provide tests of the scaling of nonthermal mechanisms of energy transport.

3. Mechanisms of Stellar Variability and the Solar Magnetic Cycle

The activity cycles of stars such as the Sun are a consequence of the periodic variation of the magnetic fields produced by the steady operation of the hydromagnetic dynamo in the convection zone beneath the visible surface. The dynamo is, in effect, an alternating-current generator with a period on the order of years. There are variations over shorter times, of course, as a consequence of the irregular eruption of magnetic flux from the dynamo up to the surface. Variation on longer time scales, such as the 70-year depression of solar activity in the seventeenth century (and also in the fifteenth century), is presumed to be a consequence of changes in the mode of convection and circulation within the star, affecting the form and efficiency of the generation of magnetic field and/or its escape to the surface. Considerable work remains to be done on the theoretical hydrodynamics of stellar convective zones before a complete picture can be obtained. The interplay between solar and stellar observations will be important to this research program, since the scaling of the period and amplitude of the resultant stellar cycles with the properties of the stellar convective zone will provide an important test of theoretical models.

4. Solar Flares and High-Energy Processes in Astrophysics

It is now recognized that a solar flare is basically a high-energy phenomenon, producing most of its luminosity in the form of x radiation and high-energy ejecta. Recently discovered high-energy phenomena in objects such as close binaries, quasars, Seyfert galaxies, radio galaxies, and the intracluster medium in clusters of galaxies all share this basic spectral characteristic. More recently, a bewildering variety of stars spread widely across the Hertzsprung-Russell diagram--RS CVn, Of, Be, dMe, W UMa, Wolf-Rayet, AO--have been shown to be sources of x-ray emission that can be most easily interpreted as coronal.

While flaring activity has been detected in only a few main-sequence stars, we can anticipate that such phenomena will eventually be shown to be nearly universal. In more extreme cases, gravitational energy must be called on to sustain an enormous luminosity that no solar-type mechanism could supply. But in terms of emission processes and basic physics, even these sources benefit from interpretation based on solar models, and there exists a whole continuum of high-energy sources from our closest stellar neighbor, Alpha Cen A, which must almost exactly resemble the Sun, to the most distant observable objects in the Universe--quasars.

5. The Interplanetary Medium

The interplanetary medium, which is the extension into space of the solar atmosphere and thus often referred to as the heliosphere, is a readily accessible laboratory for a number of physical processes important in many astrophysical systems. The solar wind is by far the best observed of the many expanding hydromagnetic flows that are important in astrophysical systems. The plasma in globular clusters and galactic halos may expand in a similar fashion; relativistic winds have been proposed for pulsars and radio galaxies; and, of course, many other stars are observed or inferred to have winds that play an important role in angular-momentum loss and, in many cases, mass loss and stellar evolution. Many propagating and convected plasma structures, such as waves, discontinuities, and shocks can be studied in situ in interplanetary space in a detail not possible in other astrophysical plasmas or, for that matter, in laboratory plasmas. It should be remembered that observations of the collisionless shock standing in the solar wind upstream from the terrestrial magnetopause represented the first detection of such a plasma structure and provided the delineation of the parametric dependence of shock structure.

As a specific example of the applicability of heliospheric plasma physics to other astrophysical systems, consider the case of cosmic-ray transport. A general theory of cosmic-ray transport has been developed, primarily in response to observations of cosmic rays in interplanetary space. This theory relates the transport coefficients to directly observable properties of the ambient plasma. The comparison of observations of both solar and galactic cosmic rays with simultaneous plasma observations is providing a stringent test of this theory. These general transport theories are applicable to cosmic rays elsewhere in the Universe, and they can be used with greater confidence and understanding because of the tests carried out in interplanetary space.

B. The Importance of Solar Physics
to the Terrestrial Sciences

To state that the Sun plays a major role in shaping the structure and determining the dynamical properties of the Earth's atmosphere, ionosphere, and magnetosphere is to state the obvious. We think it important, however, to discuss explicitly the interdisciplinary nature of solar-terrestrial physics, and thereby to demonstrate how a knowledge of phenomena of interest to each of the major disciplines--solar physics, atmospheric physics, ionospheric physics, and magnetospheric physics--is required to address major problems.

1. The Upper Atmosphere and the Magnetosphere

The terrestrial magnetosphere acts like a blunt body standing in the supersonic solar wind, with the result that a collisionless bow shock stands upstream from the magnetopause in the solar wind. Between the bow shock and the magnetopause (the boundary surface enclosing the magnetospheric cavity) lies a region called the magnetosheath, which contains shock-heated flowing plasma. Magnetosheath energy, momentum, plasma, magnetic and electric flux, and currents are coupled to the magnetosphere across the magnetopause, with the coupling rate determined by microscopic plasma processed in the thin magnetopause. It appears that the heliosphere magnetic field plays a significant and perhaps dominant role in coupling all the above quantities across the magnetopause through the magnetic-field-line reconnection process, although other transport mechanisms, such as turbulent viscosity, may be equally important.
    The solar wind stretches out the terrestrial magnetic field into a long magnetic tail, extending perhaps 1000 Earth radii downstream. The magnetic tail is divided into northern and southern lobes of opposite polarity, separated by a sheet of hot plasma that is thought to be the hot-plasma source for the inner magnetosphere. The tail and plasma sheet give rise to

the most fundamental instability of the magnetosphere--the magnetospheric substorm--in which magnetic energy accumulated in the tail is explosively converted into particle kinetic energy, with a host of important consequences: greatly enhanced auroral particle precipitation and light emission, the injection and/or acceleration of energetic electrons and ions into the inner magnetosphere and radiation belts, and the acceleration of some particles to relativistic energies in the tail. The substorm is very likely the magnetosphere's fundamental mode of energy dissipation.

The Earth's magnetosphere, ionosphere, and atmosphere are strongly coupled to one another and thereby regulate each other's time-dependent behavior. One aspect of this complex interaction is electrodynamic coupling, which refers to the coupling among flows and electric-field systems in the magnetosphere and the ionosphere-atmosphere. The second aspect of this interaction involves the escape of mass from the ionosphere into the magnetosphere, both as a thermal plasma flow (the polar wind) and as an acceleration to high energies (and subsequent escape) of ionospheric ions. The third (and potentially most far-reaching) aspect of the magnetosphere-ionosphere-atmosphere interaction is the coupling of the solar wind and magnetosphere, via the ionosphere and upper atmosphere, to the lower atmosphere. Any potential relations between solar (and solar-wind) activity and the climate and/or weather are likely to involve this last aspect of magnetosphere-ionosphere-atmosphere coupling.

2. Solar Influences on the Lower Atmosphere

The principal influence of the Sun on the lower atmosphere of the Earth is the heating produced by sunlight. The variation of climate over the past centuries has raised questions as to precisely how steady the supply of sunlight is to the upper atmosphere, and how steady that part is that penetrates to the lower atmosphere and to the surface of the Earth. Measurements of surface sunlight, carried out by the Smithsonian Institution over the first half of this century, show that there were no variations in excess of about 0.5 percent. Since variations at lower levels could have significant climatic effects, the question of the direct solar influence on climatic variations remains open.

More recently it has been discovered that there is a significant statistical correlation between drought in the western United States and the deep minimum in solar activity following alternate sunspot cycles, which occurs about once every 20-25 years. It has also been shown that the mean annual temperature in the Northern Temperate Zone, as well as the worldwide advance and retreat of glaciers, is correlated with the general level of solar activity over the centuries. Thus, the climate was cold in the seventeenth, fifteenth, and eighth centuries A.D. and in the fourth and seventh centuries B.C., when the Sun was in prolonged states of inactivity. The weather was warmer than usual in the twelfth century A.D. when the sun was hyperactive.

The mechanisms by which solar activity could affect the terrestrial climate, either by variations in atmospheric heating or by some as yet unidentified coupling to the solar wind, have not yet been established; however, a number of ideas have been proposed, ranging from variations in atmospheric electricity, to changes in the north-south mixing in the troposphere, to the nucleation of ice crystals and the formation of cirrus clouds by ions from aurorae, to the effects of ultraviolet radiation on the chemistry of ozone and nitrogen oxides in the upper atmosphere. Likewise, there are no established mechanisms that might cause the luminosity of the Sun to vary by more than 0.1 percent (expected from the coming and going of sunspots), although it has been pointed out that, if the century-long variations of solar activity are produced by changes in the general convection and circulation within the Sun, then we might expect the convective delivery of heat to the surface to vary as well. The questions raised here provide a strong link between the fundamental problem of energy transport in the Sun and terrestrial climate, thus being questions of great intrinsic scientific interest and practical importance.

C. The Role of Solar Physics in Space Physics

The interplanetary medium (or heliosphere), consisting of the solar wind and the coronal magnetic field, extends well beyond the orbit of the major planets (out to 100 AU or more). The study of the importance of winds to mass- and momentum-loss processes for stars of all classes, and of the interaction of the resultant mass flux with the interstellar medium, has become a question of central importance in astronomy. The study of the interaction of the solar wind with planetary atmospheres and magnetospheres has recently become an active subdiscipline of space physics.

1. The Interface of the Interplanetary Medium and the Interstellar Medium

As the solar wind flows away from the Sun, it must eventually be influenced by the particles and fields of the local interstellar medium. In the solar neighborhood, the interstellar medium is largely composed of four components: a magnetic field, a thermal plasma, a neutral gas, and energetic particles (Galactic cosmic rays). The magnetized thermal interstellar plasma interacts with the magnetized solar wind at a somewhat diffuse boundary separating the two plasmas (the boundary of the heliosphere), but the interstellar neutral gas and Galactic cosmic rays can readily penetrate deeply into the region of supersonic solar-wind flow.

The condition of pressure balance across the heliosphere boundary requires subsonic solar-wind flow inside the boundary and a shock transition terminating the supersonic solar wind flow somewhere upstream. The fact that the Sun is moving relative to the interstellar gas leads to a turning of the subsonic flow into a cometlike tail, distorting the shock transition as well as the heliosphere boundary. The present estimates of the minimum distances from the Sun to the shock and the heliosphere boundary are, respectively, 100 AU and 150 AU.

Galactic cosmic rays are observed directly by interplanetary space probes, and neutral atoms are observed indirectly through detection of the solar resonance-line radiation that they scatter. In both cases, the observations, together with theories describing their penetration into the heliosphere, allow one to infer the properties of cosmic rays and the neutral gas in the local interstellar medium. These inferences in fact represent the best information that we have about the neutral and energetic-particle components of the nearby interstellar medium.

2. Planetary Atmospheres and Magnetospheres

The planets' magnetospheres are cavities carved out in the solar-wind flow by their intrinsic or induced magnetic fields. Within the magnetosphere the magnetic field organizes the behavior of charged particles, plasma waves, and electric currents; it traps energetic particles to form radiation belts and confines low-energy plasma escaping into space; and, finally, it transmits hydromagnetic stresses from the magnetosphere through the partially conducting ionosphere to the upper atmosphere of the planet.

The Earth's magnetosphere was discovered in 1958. In the past several years, Pioneers 10 and 11 traversed the magnetospheres of Jupiter and Saturn, and Mariner 10 discovered an unexpected, surprisingly powerful, and highly time-variable magnetosphere at Mercury. Scaling arguments suggest that Uranus and Neptune both have large magnetospheres, and recently detected, low-frequency radio bursts from Neptune lend some credence to this suggestion.

In astrophysics, the concept of a magnetosphere has been generalized to any plasma envelope surrounding a compact central body. While every class of natural objects has its important distinguishing characteristics, the magnetospheres of tailed radio galaxies may share common features with planetary

magnetospheres; the distant portions of the magnetospheres of
pulsars and of some radio galaxies may resemble the heliosphere.
Thus, the resemblances between astrophysical and solar-system
magnetospheres could eventually motivate a unified theoretical
attack on them. Likewise, the concept of comparative magnetospheric studies should contribute to the theoretical understanding of the physical processes involved.

Most magnetospheres exhibit different variations on a common
theme: wherever Nature creates a magnetosphere, she arranges
that magnetic energy stored in it be suddenly released, accelerating a small subset of charged particles to high energy. On
the Sun, such events are called solar flares; on the Earth and
on Mercury, substorms. The electromagnetic radiation generated
by the few particles accelerated to relativistic energies might
be the dominant observable quantity in astrophysical magnetospheres; on the other hand, the energy storage and release mechanisms and the microscopic properties of the particle acceleration regions of planetary magnetospheres can be probed in situ.

IV. A PROGRAM OF SOLAR AND RELATED STELLAR PHYSICS FOR THE 1980'S

In Section II of this chapter we developed an expanded definition of solar physics as a set of related inquiries into the
phenomena displayed by matter under the influence of the
large-scale gravitational and electromagnetic fields found in
astronomical settings. We identified three major themes into
which these inquiries naturally fall and posed six problem-
oriented scientific questions related to these themes that we
believe are especially attractive for a major emphasis in the
decade of the 1980's.

The selection of the problems to be emphasized in the next
decade was strongly influenced by our perception that the new
capabilities represented by the Space Shuttle will allow the
deployment of a new class of large-facility instruments capable
of the order-of-magnitude improvements in resolution and sensitivity necessary for significant progress. A second important
observational capability is represented by deep-space probes,
such as the International Solar Polar Mission (ISPM) with its
program of out-of-the-ecliptic observations, and a close solar
encounter mission (the Star Probe) currently under study.
Techniques are now available to probe the solar interior
through detection of the neutrino flux and spectrum resulting
from thermonuclear reactions in the solar core and through
observations of global oscillations, which contain information
on the dynamics and structure of the solar interior. Furthermore, new theoretical insights and the computational capability
to investigate their consequences, which will promote interpretation of the refined observations anticipated, are now
available as well.

We present here, in an abbreviated form, the overall program that the Solar Physics Working Group believes is necessary to address the six problem-oriented questions that we wish to address. In the two following sections, we discuss this program in greater depth.

A. The Properties of the Solar Core

The study of the solar core has produced a particularly effective coupling of experiment and theory. We believe it is important to exploit fully all the available observational techniques, not only to clarify our understanding of nuclear reactions in stars, but also to test models of the composition, structure, dynamics, and evolution of stellar interiors. There are three observational approaches to this problem:

• The implementation of neutrino experiments sensitive to the low-energy neutrino flux, thus complementing the information available from the ^{37}Cl experiment (which is sensitive to high-energy neutrinos) and allowing a more stringent test of models of the composition of the solar core. Of the proposed radiochemical detectors, the ^{71}Ga experiment appears to address the fundamental issues most directly, and the Working Group strongly supports its continued development. The Working Group also urges that the ^{7}Li and ^{115}In detectors be implemented during the coming decade.

• The launch of an instrumented gravity probe that can be accurately tracked during a close (within about 3 solar radii) solar encounter, in order to determine the solar internal mass distribution and, indirectly, the core rotation rate. Although such in situ missions fall outside the area of programmatic opportunities examined by the Astronomy Survey Committee, the Star Probe mission (formerly called Solar Probe) currently under study by NASA has such a gravitational encounter as one of its four major objectives. The Star Probe mission has recently been extensively reviewed and strongly recommended by the Space Science Board's Committee on Solar and Space Physics. The Solar Physics Working Group strongly supports this mission.

• A more complete and accurate study of the global oscillation modes of the Sun, which are sensitive to the temperature and density structure of the solar core. An observational program designed to exploit the solar-core information conveyed by these oscillations will require greater sensitivity to the modes of lowest order than is furnished by present instruments. The Working Group strongly supports the development of instruments that can measure the solar velocity field with the improved accuracy required.

B. The Physics of Activity Cycles
and the Dynamical Behavior of the Convection Zone

Active phenomena in solar-type stars are consequences of dynamical processes in the convection zone. A scientific program intended to study the origin of solar activity must therefore address the question of convection-zone dynamics as well.

1. Properties and Dynamics of the Convection Zone

An exciting era of direct observational probing of the convection zone has begun. The excellent interaction between theorists and observers already established in this field is essential for further progress. This close interaction has been demonstrated by the interpretation of the properties of the normal modes of the Sun's global oscillations to obtain information on convection-zone structure and by the link between theory and observation in the study of the large-scale circulation. The major observational objectives are to define the basic properties of the convection zone, especially its depth; the rotational, magnetic, and temperature profiles with depth and latitude; the wide range of eddy sizes as a function of depth; and the temporal variation of these properties. It is also necessary to determine the nature of large-scale circulation and temporal and spatial variations of differential rotation.

The observational programs required are (a) the determination of the total solar velocity field with great precision (about 1-10 km/sec) for extended periods (greater than 30 days) and, if possible, with stereoscopic capability; (b) precision differential radiometry (to 1 part in 10^5) of the convective cell structure of the photosphere; and (c) measurement of the large-scale solar magnetic field with a sensitivity of 1 gauss.

2. The Solar-Activity Cycle

The solar-activity cycle produces extensive variations in what might be called the "solar outputs." These include systematic changes in the spectrum and flux of electromagnetic radiation and of charged particles; variations in the interplanetary magnetic field; and changes in the composition, mass flux, and energy flux of the solar wind. It is the variation of the magnetic structure of the convection zone (the ultimate cause of the active phenomena in the solar atmosphere) that underlies these manifestations of the activity cycle. The operation of the solar dynamo, which is responsible for the solar magnetic field, changes in response to variations in the transport of energy within the convection zone. Therefore, the study of the solar activity cycle involves two complementary observational programs. First, we must study the variations in the structure and dynamical behavior of the convection zone over a full magnetic cycle (22 years). We must simultaneously measure the variations of the solar outputs accurately over a complete magnetic cycle in order to understand (a) their connection to the variations in the basic parameters of the convection zone, chromosphere, and corona and (b) their influence on the Earth's environment. The accuracy with which these measurements must be made will require significant improvements to the stability and absolute precision of previous observational capabilities.

3. The Physics of Stellar Activity and Atmospheres

During the next decade, major advances in our understanding of stellar-dynamo processes and resulting stellar activity will require a determination of the functional dependence of dynamo efficiency on critical atmospheric parameters. More explicitly, we must obtain observations that reveal the dependence of the period and amplitude of activity cycles on stellar rotation, gravity, depth of the convection zone (which depends on effective temperature and radius), and the internal angular momentum distribution, which itself presumably depends on stellar age and the strength of the stellar wind and surface magnetic field.

These are difficult observational questions, but they can be addressed through synoptic observing programs employing high-resolution spectroscopy from space and from the ground. It is important to determine whether stellar activity is manifested mainly in the number and size of starspot and plage regions on a star or whether the enhanced magnetic fields near activity maximum are more uniformly distributed across the surface of a star. Another critical question is the statistical probability and typical duration of Maunder Minimum-like absences of activity in stars: Which stars undergo this behavior, and why? With the insight gained from these observational results, the continued theoretical study of dynamo processes should become still more productive.

4. Programmatic Implications

A systematic program for the study of the solar cycle and the dynamics of the convection zone has been developed by the Solar Cycle and Dynamics Mission (SCADM) Working Group of NASA in a recent report. This program can be largely carried out within the ongoing programs of NSF and NASA, provided the level-of-effort budget of the Explorer program is substantially

augmented. The principal components of the Solar Cycle and
Dynamics Program are as follows:

* A major effort in the theoretical modeling of solar
interior dynamics.
* The development of sensitive techniques for the observation of the solar velocity field, brightness variations,
diameter variations, and the large-scale magnetic field,
together with a program of long-term (about 1 month) observations using these techniques, initially from the South Pole
of the Earth and subsequently from the Space Shuttle in Sun-
synchronous orbit.
* The extension of this observational program to longer-
duration observations (greater than 6 months) with a complement
of instruments on a free-flying Solar Interior Dynamics Misson
(SIDM), which might be implemented through the Explorer Program
or on a space platform. In addition to the study of the structure and dynamics of the convection zone, the SIDM will carry
out high-precision measurements of total solar radiative flux
and the solar-spectral irradiance. It is important, however,
that the program of high-precision solar-flux measurements
begun by the Solar Maximum Mission be continued through Shuttle
and other flight opportunities until SIDM becomes operational.
* A Solar Coronal Explorer (SCE) mission, which will
study the three-dimensional structure and evolution of the
corona and solar wind in conjunction with in situ observations
by the ISPM and the Interplanetary Laboratory (IPL) of the
Origin of Plasmas in the Earth's Neighborhood (OPEN) program.
* The extension and refinement of observational programs
initiated by the SIDM and SCE missions through the observational programs of the Solar Optical Telescope and the other
Shuttle facilities, the Advanced Solar Observatory, and the
Star Probe Mission.

The major theoretical and developmental programs required
for the Solar Cycle and Dynamics Program and supporting observational programs will be carried out in large measure with
major ground-based instrumentation and computational facilities
at the Sacramento Peak Observatory (SAO), at Kitt Peak National
Observatory (KPNO), at the Very Large Array (VLA), and at the
High Altitude Observatory.
A second major aspect of the study of activity cycles is
the observation of activity in other stars. The study of stellar activity cycles through the observation of chromospheric
line profiles with ground-based telescopes has already yielded
significant results (most notably with the 1.5-m telescope at
the Mt. Wilson Observatory).
The Working Group believes that this activity would be
greatly enhanced by the implementation of a dedicated stellar
telescope of moderate aperture (2.5 m) for synoptic observa-

tions of stars to be operated in conjunction with the solar
programs at the National Astronomy Centers. We also note that
several proposed future x-ray observatories in space and
several approved or proposed Explorer missions in the areas of
astronomy and astrophysics (e.g., the Extreme Ultraviolet
Explorer and a Stellar Coronal Explorer) will have observations
of stellar activity as major or principal objectives.

5. Impact of Variability on the Terrestrial Environment

Although the scientific questions raised here are not strictly
astrophysical, the interaction between solar observations and
observations of the short-term and long-term variation in the
terrestrial environment has been fruitful for both astrophysics
and geophysics, and it promises to become more so in the
immediate future. The discovery of the correlation of the level
of solar activity with extremes of the terrestrial climate is
only one example of this interaction. Such correlations are
important astrophysical tools for the study of solar-activity
cycles. Solar physics must be an active participant in this
collaborative program. This will require a vigorous exploitation of, and sensitivity to, the opportunities presented by the
proxy record of variations in the solar wind and the solar
radiative flux as represented in, for example, lunar samples
and biological and geologic records. It will also require a
close collaboration between solar observations and magnetospheric, ionospheric, and atmospheric observational programs
such as those of the OPEN mission and the Solar Terrestrial
Observatory (STO), which explicitly envision a full complement
of related solar, magnetospheric, ionospheric, and atmospheric
observations.

C. Nonradiative Heating and Transient Phenomena
in the Solar Atmosphere

We presented above a program directed toward the study of the
physical processes operating in the solar convection zone, as
well as the manifestation of these and similar processes
operating in other stars in the variation of solar and stellar
outputs. The second aspect of the study of stellar-active
phenomena is the study of the structure and dynamical behavior
of the extended solar atmosphere itself and of similar phenomena in other stars.
The discoveries of the past decade have laid the foundations for two major approaches to the problem of the structure
and heating of stellar atmospheres; one makes use of detailed
observations of the solar atmosphere, whereas the second makes
use of comparative observations of the atmospheres of stars

with a wide range of fundamental properties. These approaches
are highly complementary to each other, as discussed below.

In considering transient and high-energy phenomena in the
solar atmosphere, we take the basic physical mechanisms responsible for metastable energy storage, release, and deposition in
the solar atmosphere as the focus of our discussion. The chain
of events exhibiting these essential phases occurs in many
astrophysical and space-physics situations and is called a
flare.

Among the highest priorities for the early part of the next
decade must be the understanding of the data from the Solar
Maximum Mission (SMM). We anticipate that major advances in
our understanding of flares, especially the impulsive energy
release, will result from the complete analysis of SMM observations. However, we know that many of the fundamental physical
processes that are important in flares occur in structures on a
scale too small to be resolved by SMM instruments; these will
require a new generation of facility-class instruments for
their study.

1. The Solar Atmosphere

Earlier observational programs on the Orbiting Solar Observatory
satellites, Apollo Telescope Mount, and SMM have demonstrated
that the basic plasma processes governing the propagation and
transformation of energy in the solar atmosphere occur on
physical scales that are beyond the resolving power of present
instruments. With the advent of the Space Shuttle, we are now
in a position to develop a new generation of high-resolution
solar telescopes covering optical, ultraviolet, extreme-
ultraviolet, and soft x-ray wavelengths, which we believe will
allow us to resolve the structures in which at least some of
these fundamental processes occur.

The first of these new facilities, the Solar Optical
Telescope (SOT), is already under development. The SOT is a
1.25-m aperture, diffraction-limited telescope designed to
achieve a tenfold improvement in angular resolution (about
0.05-0.1 arcsec) by comparison with previous observations.
Comparable improvements in resolution are anticipated at x-ray
and EUV wavelengths for instruments now under study by NASA.
The observations anticipated from this new generation of instruments will allow, for the first time, the comparison of
detailed, self-consistent physical models with observations of
the structures (including sunspots, magnetic flux tubes in the
chromospheric network and in plages, spicules, and coronal
bright points) that dominate the solar photosphere, chromosphere, transition region, and corona.

2. Stellar Atmospheres

The second major approach to the structure and dynamical behavior of stellar atmospheres involves the study of the properties of stellar chromospheres and coronas, now known to exist in stars with a wide range of fundamental parameters (e.g., mass, temperature, age, luminosity, and rotation rate). The observation of stellar coronas and chromospheres will allow the development of models that relate fundamental stellar properties to the character of the active phenomena associated with the star. Despite the necessary simplicity of these stellar models, it is important to estimate the nonradiative heating rates in stellar models with a range of parameters to determine the role that each parameter plays in overall atmospheric structure and behavior.

3. Metastable Energy Storage

The major determinant of the energy source and initiation of the flare instability is the detailed structure of the coronal magnetic field. The current that flows in this field provides the free-energy supply that drives the flare release, and its spatial configuration (including plasma profiles) specifies the singular flare site.

In order to understand the preflare state, vector-magneto-gram measurements must be made from space with a resolution of less than 0.1 arcsec, and an improved generation of field-calculation algorithms will have to be developed to specify the coronal field constructed in accordance with these boundary values. A determination of the thermal structure and evolution of the preflare state requires coordinated, mutually aligned, spectroscopic measurements from the optical region of the spectrum through x rays. The spatial resolution and mutual registration should be better than 0.1 arcsec (72 km), with a sample rate up to one per minute available.

4. Magnetic Reconnection and Particle Acceleration

This topic is the least understood aspect of plasma and high-energy astrophysics; it defines the transformation of slow, large-scale motion and magnetic stress into impulsive, ener-getic, small-scale phenomena.

It is now believed that such "flare" events arise from the relaxation of sheared magnetic fields, subsequent to the recon-nection from one frozen-in state to another in which the cur-rents are filamented and can quickly decay. The major initial product of the reconnection process (in which the magnetic fields both vary and move) is a selected population of acceler-ated particles.

To relate the poorly resolved magnetic-field source and the imprecisely determined output radiations, we currently rely upon theory. The required calculations are nonlinear in their time development and coupled in their spatial geometry. They must include the plasma and field dynamics, together with heating and energy flow. They must also encompass kinetic effects in the acceleration and microwave radiation processes and a proper treatment of the bremsstrahlung targets, all connected by nonlocal propagation effects confined by the magnetic field. Beyond an expanded theory program directed to these ends, we must emphasize the observations needed to clarify the impulsive phase, keeping in mind the advances achieved by SMM.

A primary objective is the development of better methods of measuring the accelerating electric field directly as a function of position through the Stark effect. The next step would involve correlated, high-resolution observations of hard x-ray and microwave bursts. (The hard x-ray observations must have subarcsecond resolution.) It is also critical to understand the relationship of the acceleration of electrons to the acceleration of the solar cosmic radiation, both by direct observation of these particles and by highly spectrally resolved gamma-ray observations. Again, high-resolution observations of hard x rays will also play a key role.

5. Energy Transformation and Deposition

We here focus attention on the high-temperature phenomena caused by the dissipation of the explosive flare event. The relativistic electron population resulting from the initial flare acceleration, described above, is mostly confined in closed magnetic structures embedded in the solar surface. In this environment, the electrons heat the ambient plasma and especially the chromospheric end points, which then react and drive material and radiation back into the corona. Then the decay phase of the flare begins, and energy is steadily lost by x radiation and by thermal conduction to the chromosphere and subsequent further EUV emission.

Further clarification of this scheme, which has just begun to become apparent as a result of recent Skylab data analyses and SMM results, requires a complex set of coordinated observations. As for other coronal phenomena, the major needs are spatially resolved (less than 0.1 arcsec) determinations of plasma number density and temperature from spectral lines ranging up to soft x rays, with 1-10 sec resolution, and measurements of the vector magnetic field. Having started with the field, we end with the field, so that the postflare magnetic energy can be determined. This residue, along with the original supply and measured outputs listed above, will allow the verification of the complete flare energy budget.

6. Programmatic Implications

In order to carry out the scientific program presented above, several major new observational and theoretical programs are required:

* A comprehensive program of theory and modeling, with the principal aim of improving our ability to interpret solar observations in terms of the physical parameters of the solar plasma and to relate these parameters to underlying plasma instabilities. An important component of this program is the modeling of the basic plasma processes that underlie the flare mechanism.

* The principal observational tool required is a comprehensive ensemble of Shuttleborne facility-class and Principal Investigator-class instruments, built around the SOT, which are capable of simultaneous, high-resolution observations covering the full spectral range from x rays to the near infrared region. Once developed, this ensemble should be placed upon a free-flying satellite or space platform to form an Advanced Solar Observatory (ASO) for the long-term study of the physics of solar and stellar activity. For the study of flares, it is critical that the initial instrument complement of the ASO, which might include the SOT, the Solar Soft X-Ray Telescope Facility, the EUV Telescope Facility, and a Pinhole/Occulter Facility, which will allow high-resolution (about 0.1 arcsec) hard x-ray observations and high-resolution observations of the corona and coronal transients, be augmented by a high-resolution gamma-ray spectrometer in time for the 1991 solar maximum. In addition to the space observations by ASO, a program of ground-based observations emphasizing the observation of solar magnetic fields with improved vector magnetographs (at SPO and KPNO) and of microwave bursts (with the VLA) is critical to the flare-oriented scientific program that we have outlined.

* A second major observational capability is presented by a set of optical, EUV, and x-ray imaging instruments currently under study as a major component of the Star Probe mission. These instruments will provide a unique opportunity to observe the structure of the solar atmosphere with ultra-high resolution (about 7-10 km), which is 3-10 times higher than can be achieved from Earth orbit even with the ASO.

* The development of strong observational programs, from both the ground and space, for the study of the coronas and chromospheres of other stars.

* In addition to these new programs, the continued development of new and more sensitive techniques for optical and radio observations of the solar atmosphere as part of the ongoing programs of the National Astronomy Centers and efforts supported by the NSF grants program is critical to continued progress in this area.

D. Properties of the Corona and Interplanetary Medium

There are three principal questions in coronal and heliospheric physics that can be effectively addressed in the decade of the 1980's and the early 1990's:

• What factors control the mass and energy balance of the corona, and what are the implications of these factors for the coronal temperature, coronal expansion (the solar wind), and the transport of angular momentum away from the Sun?
• What is the three-dimensional structure of the corona and heliosphere, how does this structure evolve with the solar activity cycle, and how does the impulsive ejection of mass by coronal transients (which supplies a significant fraction of the mass of the solar wind) affect this structure?
• What are the physical processes responsible for the acceleration of the solar wind, and what mechanisms are responsible for the large variation in the observed solar-wind composition?

All three of these problems can be addressed by a coordinated program of high spectral and spatial resolution observations of the chromosphere and corona, together with in situ observations of the heliospheric plasma in three dimensions and as close to the Sun as possible. The development of models of energy and mass flow in the solar wind, making use of empirical data to specify the three-dimensional structure of the corona and heliosphere, must proceed in concert with the observational programs. Other important theoretical problems include the development of magnetohydrohynamic models of the basic processes responsible for driving the transients and for the energetic-particle acceleration associated with transients (in particular, with flare-related transients). Our main conclusions and concerns are as follows:

1. During the next decade we shall, with ISPM, have a unique opportunity to study the three-dimensional structure of the solar wind and interplanetary magnetic field, and the influence of this structure on the interplanetary modulation of Galactic cosmic rays. (As noted near the beginning of this chapter, the Working Group believes that the implementation of ISPM as a dual-spacecraft mission with imaging capability is essential to an orderly and effective program of solar and heliospheric studies.) A comprehensive understanding of coronal and heliospheric structure and dynamics will, however, require complementary in-ecliptic observations by a Solar Coronal Explorer and the Interplanetary Laboratory of OPEN. With the same spacecraft and with some careful planning, it may be possible to gain new information about the solar-wind transport of angular momentum and thus new insight into the problem of solar and stellar spindown.

2. The major scientific objectives of heliospheric physics, which will be greatly furthered by the observational programs of ISPM, SCE and IPL, can, however, only be fully addressed by two advanced missions:

(a) A close-encounter mission (the Star Probe), which will allow in situ observations in the region of solar-wind acceleration and a unique view at ultra-high resolution (7-10 km on the Sun, or the equivalent of 0.01 arcsec from Earth orbit) of coronal and chromospheric structure; and

(b) Synoptic studies of the evolution of the chromospheric and coronal structures and dynamical events that control the large-scale structure of the heliosphere with the assembly of high-resolution instruments, especially the Pinhole/Occulter Facility, on the ASO. In this connection, the ASO should include radio spectrographs and polarimeters operating from decametric to kilometric wavelengths for the study of particle streams, collisionless shock waves, and shock-induced acceleration in the corona and solar wind.

V. SCIENTIFIC OPPORTUNITIES FOR THE 1980'S

The major scientific problems identified earlier are discussed here in more detail under the following headings: the solar interior; the convection zone and solar activity; nonthermal and dynamical phenomena in the solar atmosphere; and the dynamics of the heliosphere.

A. The Solar Interior

The general characteristics of the Sun's interior structure can be deduced from theoretical models, with observed values of the solar radius, luminosity, mass, and surface rotation rate as boundary conditions. Such interior models (illustrated by Figure 1.1) can give a satisfactory account of the general evolutionary behavior of the Sun but cannot resolve a number of important questions concerning the detailed composition, temperature, density, and rotational profiles of the interior; nor can they predict such phenomena as the activity cycle and the structure of the chromosphere and the corona. The detection of the solar-neutrino flux marked the beginning of a scientific program that will ultimately allow detailed comparisons between sophisticated models of the solar interior and direct measurements of the solar-neutrino flux. The use of two other techniques that probe the solar interior--the study of the global oscillations of the Sun and the observation of the perturbations caused by the solar quadrupole moment on a close gravitational probe--will become possible in the 1980's. The theoretical analysis required to interpret these observational techniques is well advanced.

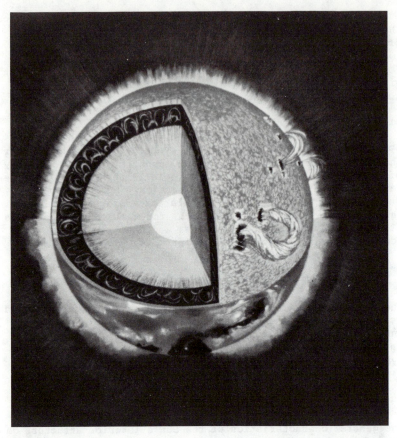

FIGURE 1.1 A cross section of the Sun, indicating the inner, energy-generating core of about 0.2 solar radius; the radiative envelope, which extends out to about 0.8 solar radius; and the outer convective shell.

1. Core Structure and Nuclear Reaction Rate

The solar luminosity is due to two thermonuclear reaction chains that convert hydrogen into helium in the solar core. The main reaction chain is the proton-proton (p-p) chain, in which protons interact to build up ^{2}H and ^{3}He; the latter is then burned to ^{4}He in one of three branch reaction chains. A small fraction of the neutrino flux (about 1 percent for standard models) arises from reactions in the CNO cycle, in which a series of reactions involving isotopes of C, N, and O results in the conversion of 4 protons into ^{4}He. The rela- tive flux of the neutrinos from the various branch reactions is

sensitive to such parameters of the solar model as relative
initial abundances and the extent of mixing between the core
and envelope during the Sun's lifetime. The solar-neutrino
spectrum predicted by the standard solar model is shown in
Figure 1.2; clearly, observations with radiochemical neutrino
detectors having selective energy response (i.e., neutrino
spectroscopy) can provide a measure of the rates of the various
major reactions.

The event rate for the ^{37}Cl experiment is dominated by
the high-energy neutrino flux from the p-p branch reactions in
which ^{8}B is built up from ^{3}He and ^{4}He via ^{7}Be and proton
capture, followed by positron and neutrino emission to yield an
excited state of ^{8}Be, which in turn decays into two ^{4}He nuclei;
these branch reactions account for less than 1 percent of the
overall p-p chain-reaction rate. The neutrino detection rate
predicted by the standard solar model for such high-energy
neutrinos is about 5 ± 1.4 SNU (the Solar Neutrino Unit being

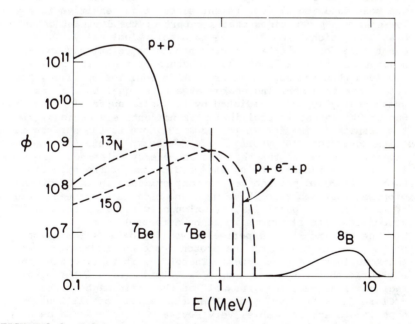

FIGURE 1.2   Solar energy neutrino spectrum, from a 1971 model of
J. Bahcall and R. Ulrich.   Solid lines: p-p chain neutrinos;
broken lines: CN cycle neutrinos.   Fluxes are in units of
number of neutrinos/cm^2/sec/MeV for continuum sources and
number of neutrinos/cm^2/sec for line sources.

defined as one event per 10^{36} atoms of ^{37}Cl per second).
The measured rate of 1.8 $\pm$ 0.4 SNU is in serious disagreement
with the predictions of the standard model; furthermore, recent
calculations utilizing the most recent laboratory opacity data
suggest that the predicted rate should be even higher, some 7
SNU. A number of modified solar models employing such assumptions as lower primordial abundances of He, C, N, O, and other
elements in the core, or frequent, complete mixing of the
interior, have shown that the predicted ^8B neutrino flux can
in these cases be reduced sufficiently to produce overlap of
the errors of the experimental and theoretical results. A
second possibility is that modifications to the fundamental
theory of the neutrino are necessary; significant, but
certainly not yet definitive evidence for this possibility has
recently been advanced by physicists.

Observations with three additional radiochemical detectors--
^{71}Ga, ^7Li, and ^{115}In--could provide a fairly complete picture
of the spectral distribution of the solar-neutrino flux, and
therefore of the relative reaction rates for the main thermonuclear reaction branches. The ^{71}Ga detector would provide
the most fundamental measurement since it is sensitive to the
low-energy p-p neutrinos that are part of the dominant chain of
branch reactions providing the main solar luminosity. The
event rate for the ^{71}Ga detector should be relatively independent of the details of the structure of the solar core,
provided that the Sun is at present in equilibrium (i.e., the
solar core is generating energy at a rate equal to that at
which energy is being radiated by the solar surface). That the
Sun is at present in equilibrium is not entirely obvious, since
the transport time for energy from the core to the surface is
quite long (some 10^6 years). If, for example, the ^{71}Ga
experiment demonstrates that the low-energy solar-neutrino flux
is also significantly below the predictions of the standard
theory, it could provide very strong evidence in favor of the
proposal that neutrinos have mass and oscillate between three
or more kinds of particles, of which only those neutrinos
associated with electrons would be detectable.

The ^7Li and ^{115}In experiments would provide information
on the $^7Be/^7Li$ branch of the proton cycle and CNO-cycle
thermonuclear reaction rates. Therefore, it is important that
a program that envisions the implementation of all three new
radiochemical detectors, as well as the refinement of the
^{37}Cl results, be developed. When the results of all four
experiments are available, very severe constraints can be
placed on the abundances and thermal structure of the solar
core; questions on primordial solar abundances and the extent
of interior mixing can then be addressed.

2. Core Structure and the Solar Quadrupole Moment

The presence of a solar quadrupole moment is a consequence of the breaking of spherical symmetry by the solar rotation. Preliminary analysis of the solar p-mode oscillations indicates that the rotation rate of the convection zone increases with depth; a rapidly rotating solar core is therefore a distinct possibility. Just as the present uncertainty about abundances in the solar core introduces considerable latitude in the details of solar models, so too does our lack of knowledge of the rotational and magnetic profiles with depth. Therefore, models constructed to explain the low counting rate of the ^{37}Cl experiment through inclusion of a rapidly spinning core or a large central magnetic field, and which thus alter the pressure and temperature profiles of the interior, cannot be ruled out at present. For such models, and for models that postulate a higher hydrogen abundance in the core (either because of primordial inhomogenetics in the young Sun or because of substantial mixing of the interior), the resultant variation of density with radius is different from that assumed in the standard model and thus corresponds to a different quadrupole moment. For example, the standard solar model predicts a quadrupole moment (J_2) of 1×10^{-7}; a model with a well-mixed core, but otherwise like the standard model, yields a J_2 value of 2×10^{-7}. A model with inner mixing, a convective core, and a magnetic interior that rotates at twice the surface rate predicts a J_2 value of 7×10^{-7}. A close gravitational probe, with a drag-free system and accurate tracking, can measure J_2 with an accuracy of 1 part in 10^8; such a probe would thus permit a clear discrimination among interior models with different core parameters. The measurement of the solar quadrupole moment furthermore complements information derived from neutrino spectroscopy, since the value of J_2 is more strongly dependent upon the interior rotational and magnetic structure than on the neutrino detection rate.

3. Solar Oscillations and Interior Structure

As we shall discuss shortly, the well-known 5-min oscillations of the solar atmosphere have been shown to be due to higher spherical harmonics of the nonradial p-mode (pressure) acoustic oscillations of the convection zone. Although they have not yet been detected, low-order p-mode oscillations that involve the keep interior should also be present. Furthermore, a number of observers have claimed the detection of oscillations of longer periods (from about 10 to 160 min), which they suggest are low spherical-harmonic radial g-mode (gravity) oscillations that involve the entire Sun. Both of these oscillation modes can be used to probe the interior structure of the

Sun. The observational programs intended to detect these oscillations make use of precise measurements either of the solar diameter or of the velocities of widely separated areas on the solar surface. The amplitude of these oscillations is small, and the brevity of the data records available makes it difficult to eliminate completely various sources of noise; nevertheless, it appears that an increasing number of observers and theorists regard the detection of both low-order p and g oscillation modes as possible and as providing a new observational window for the study of the solar interior. In addition to their potential for the study of the solar core, the solar radial and nonradial oscillations provide a technique that can probe the depth of the convective zone and investigate a possible shear boundary layer between a magnetic, rapidly rotating core and the convection zone. The existence of such a turbulent shear layer could have important consequences for mixing of the outer layers of the Sun and the thermonuclear core. It is imperative that techniques of higher sensitivity be developed to confirm and exploit the opportunities presented by low-order solar oscillations.

More recently, some observers have suggested that there are indications of oscillations with very long periods (about 12 days); some theoretical analyses suggest that oscillations with periods as long as 10 years may be present. It is certainly too early to judge these claims; however, it is clear that the 1980's will be an exciting period of exploration with respect to the fundamental oscillation modes of the Sun.

B. The Convection Zone and Solar Activity

It is the interaction between rotation and the large- and small-scale motions generated by the convective transport of energy from the solar interior to the surface that causes the solar magnetic field to vary and underlies the solar activity cycle. Therefore, the scientific questions posed by the rotational structure of the Sun's convective envelope, large-scale circulation, the convective transport of energy, and the solar dynamo are closely linked to the solar activity cycle and are most effectively studied through a single coordinated program.

Four recent developments have shaped scientific opportunities for the study of the structure and dynamical behavior of the convection zone and of solar activity. These developments are as follows:

• The availability of computers of sufficient capacity to permit hydrodynamic models of convection and circulation in the Sun to be constructed, allowing direct comparison between experiment and theory.

• The interpretation of the 5-min oscillations as global
p-mode oscillations in the convection zone. The properties of
these oscillations can be used as a probe of convection-zone
structure and dynamics, provided sufficiently precise records
of surface velocity fields can be obtained.
• The discovery of the intimate connection between the
large-scale structure of the solar magnetic field and the
structure and energy balance of the corona and heliosphere.
• The discovery of activity cycles in other main-sequence
stars. The study of stellar activity will allow us to determine
how the nature and intensity of activity cycles and other manifestations of stellar activity (such as winds and flares) depend
on such basic stellar parameters as age, mass, surface temperature, and radius.

A program for the study of convection-zone structure and
dynamics and the solar cycle should, therefore, have five major
components:

• A theoretical program of modeling of the hydrodynamic
structure of the convection zone, of the solar dynamo, and of
the p-mode oscillations;
• An observational program for the study of the velocity
field and temperature structure of the solar surface;
• An observational program to study the evolution of the
large-scale structure of the corona;
• An observation program to study the variation of solar
radiative and particulate flux over the full solar cycle; and
• Observational and theoretical programs for the study of
activity cycles in other stars.

1. Major Problems in Convection Studies

Convection is a basic mechanism of energy transport in all cool
stars; it drives the global circulation and indirectly (through
the dynamo mechanism) generates their magnetic activity cycles,
therefore providing the nonthermal energy that heats their
chromospheres and coronas and accelerates stellar winds.
Models of convection in the Sun must explain the three
scales of convective motion observed in the photosphere:
granulation cells (diameter 1500 km), supergranulation cells
(diameter about 30,000 km) and "giant" cells (diameter about
300,000 km). (The solar diameter is 1.4×10^6 km.)
Considerable progress has been made in the past decade in
modeling large-scale convection (the giant cells). The most
advanced models consider nonlinear axisymmetric convection in a
deep rotating spherical shell; however, smaller scales of convection and turbulence enter the theory only as parameterized
coefficients, and variations in density (covering a factor of

10^5 within the convection zone) are not rigorously treated. These models have shown how the solar convection drives the latitude-dependent rotation of the photosphere and predict a meridional flow that agrees in direction (if not in magnitude) with the observed flow. However, the approximations employed result in a predicted pole-equator temperature difference much larger than that observed.

Thus, at present, there are two main problems in solar convection: (a) to explain the existence of three preferred spatial scales at the surface and to understand the depth variation of the spectrum of scales and (b) to explain the observed latitude variation of rotation in the face of a uniform surface radiative flux. In the future, two additional problems must be addressed: (c) to explain the gross regularities observed in the 11-year magnetic cycle of the Sun (Hale's laws of polarity and Sporer's "butterfly" diagram of sunspot frequencies--see Figure 1.3); and (d) to understand how the convection and circulation can jump between two bistable modes, as evidenced by the existence of large fluctuations in solar activity during the past (e.g., the Maunder Minimum).

Further progress in modeling convection will require a significant extension of present-day techniques to permit the development of a theory of compressible, nonlinear, nonaxisymmetric convection in a deep rotating shell. Present-day computers like the Cray 1 are adequate to the task if sufficient time is available on them. However, new theoretical developments are needed to treat convective scales that are smaller than giant cells and turbulence.

Several kinds of observations are needed to guide the theory. Doppler velocity measurements of the giant cells and the surface-brightness fluctuations in supergranulation and giant cells are most important. Giant cells are expected to have lifetimes of several months, horizontal speeds of at most 10 m/sec, and extremely small temperature fluctuations. To observe them, we need to develop and apply highly stable velocity detectors, such as the Fourier tachometer currently under development at Sacramento Peak Observatory. Such measurements might be carried out effectively from a station in Antarctica, where the Sun is continuously visible for periods of several days during the few months of an austral summer.

Defining the kinematic properties of granules depends on obtaining observations of higher resolution (better than 0.1 arcsec) for extended periods of time. In order to study convective processes operating at the smallest scales, we will require the high-resolution capabilities of SOT and, ultimately, the very high resolution that only the Star Probe encounter can provide.

39

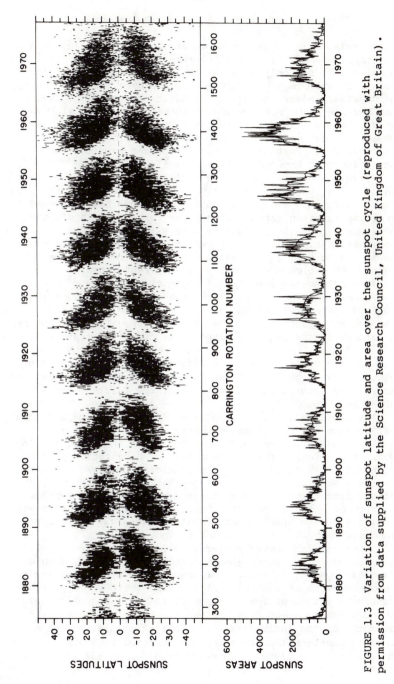

FIGURE 1.3 Variation of sunspot latitude and area over the sunspot cycle (reproduced with permission from data supplied by the Science Research Council, United Kingdom of Great Britain).

2. Circulation, Rotation, Pulsations

Large-scale motions, including rotation, are vital features of all stars. An understanding of large-scale motions of the Sun is essential, not only because the Sun is the only star in which such motions can be studied in adequate detail to test our physical understanding of the basic processes involved but also because these motions are intimately connected with solar activity and possibly with long-term fluctuations of solar properties that may have direct consequences for the terrestrial climate. Further progress in solar-dynamo modeling requires better knowledge of large-scale motions.

During the past decade, the differential rotation of the Sun has been studied extensively both from the ground and from space. Virtually every observable phenomenon has been used to determine rotation rates, which prove to have a disquietingly large range of values. Some of this variation is due to a solar cycle association; additional variation has been explained by the anchoring of various features to different depths below the photosphere, which are rotating at different rates.

An exciting new tool for studying the rotation of the photosphere and the upper layers of the convection zone arose from the discovery that the 5-min oscillation is due to a standing-wave pattern in the solar interior. The rotation rate at various depths can be derived by tracking the transverse motion of various temporal and spatial components of the oscillatory pattern as illustrated in Figure 1.4. The variation of rotation rate with depth found from this technique suggests an increase of several percent inward for the first 15,000 km below the photosphere.

The fluid velocities associated with circulation are much smaller than the rotational velocity and the velocities of smaller-scale phenomena such as granulation and supergranulation. Accordingly, the determination of circulation patterns is one of the most difficult observational problems in solar physics. Recently, several independent spectroscopic investigations have demonstrated a poleward meridional flow with a velocity of about 30 m/sec; unfortunately, this result is very sensitive to the spectral-line shifts that occur as observations are made closer to the solar limb.

It has so far proved extremely difficult to find Doppler-shift evidence for other global circulation effects, such as giant cell patterns, because of obscuration by smaller-scale motions as well as instrumental and limb-shift effects. The transverse motion of features associated with magnetic fields has provided some evidence for the existence of giant cells. Estimates of the latitudinal transport of angular momentum by presumed giant-cell motions indicate transport toward the equator in the sense required to maintain differential rotation. In spite of the difficulty of the observations, recent instru-

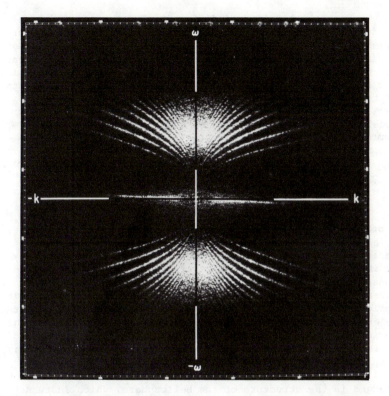

FIGURE 1.4 Properties of coherent, global 5-min oscillations
of the Sun as a function of k and ω, in which k is the
wavenumber of the oscillating element and ω is the frequency.
(Courtesy of J. Harvey, E. Rhodes, and T. Duvall.)

mental advances suggest that a major observational effort
should be directed toward this important problem.

Large-scale motions are expected to be accompanied by small
variations in radiative flux. These flux-intensity patterns
might therefore contribute to changes in solar luminosity.
Spatially resolved measurements of brightness fluctuations have
not yet shown any evidence for the existence of radiative flux
patterns associated with large-scale motions. Full-Sun measurements have suggested slight changes in the apparent spectroscopic temperature of the Sun that could be associated with
large-scale motions. It is important to pursue these measurements, since they could yield information about temperature
structure with depth in giant cells and provide another measure
of subsurface rotation.

Theoretical modeling of differential rotation and circulation has interacted well with observational advances. For example, the observed surface poleward meridional circulation is in accord with the predictions of one model but is contrary to that of another. A polar vortex was predicted by one model but not observed; this result led to an increase in the depth of the model convection zone as required by evidence from oscillation probing of the convection-zone depth. Future modeling efforts should focus on more realistic physical representations of the Sun, especially by incorporating compressibility and through coupling of rotation, circulation, and convection.

3. The Solar Activity Cycle

A broad variety of phenomena--solar, heliospheric, and terrestrial--show quasi-periodic variations correlated with the variation of the mean sunspot number. The essential characteristics of the activity cycle are the variation of the sunspot number and the migration of sunspots from higher to lower latitude (as shown in Figure 1.3), which have an 11-year period, and the magnetic polarity of sunspot groups and of the polar fields of the Sun, which reverse after each cycle. The period of the cycle is therefore more correctly 22 years, and it is in fact a magnetic cycle. The other phenomena associated with the activity cycle--the waxing and waning of the corona (and solar EUV and x-ray flux) and of the solar wind, and the changes in the structure of the heliosphere, which in turn modulate the transport of Galactic cosmic rays--are all a consequence of the variation of the solar magnetic field, as are the characteristics of the sunspots themselves.

No two solar cycles have been observed to be identical in the past few hundred years. Furthermore, recent historical research has shown that solar activity can be depressed for periods of decades. Figure 1.5 illustrates this phenomenon, showing the abnormally low sunspot numbers observed in the late seventeenth century and documented by Maunder through the examination of contemporary records. Because the Maunder Minimum occurred shortly after systematic records of sunspot counts were begun, the importance of this phenomenon has only recently been appreciated. We now know that solar activity has been either abnormally high or abnormally low about 1 percent of the time during the last 70 centuries. The mean annual temperature in the Northern Temperate Zone has tended to follow levels of solar activity; we will return to this connection between solar activity and climate in a later section.

A program for the study of the solar activity cycle should have two aspects. First, the manifestations of the cycle should be carefully studied with much greater precision than

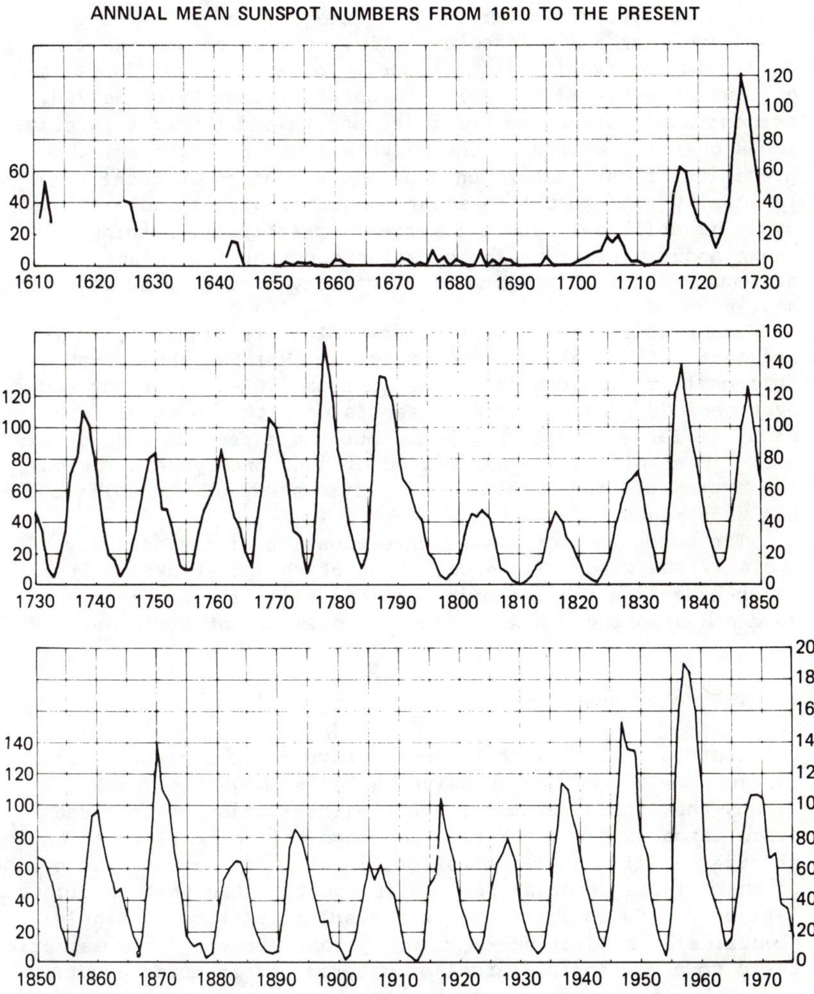

ANNUAL MEAN SUNSPOT NUMBERS FROM 1610 TO THE PRESENT

FIGURE 1.5 The Maunder Minimum in solar activity, as measured
by sunspot number, during the approximate period 1635-1705
(courtesy of J. Eddy).

has been possible in the past. While we know that the solar
flux shortward of about 4000-Å wavelength shows considerable
variations over a solar cycle, the magnitude of this variation
and its time variation are not well known. Whether the
integrated solar luminosity varies significantly with the level
of solar activity is also not well known at present. The
variation of the large-scale structure of the corona and of the

solar wind over the solar cycle is complex and not well understood even for a single solar cycle. These are important and difficult observational questions. For example, it is estimated that a sustained reduction in solar luminosity of only 0.5 percent would drive the polar ice sheets and temperature climate zones over 160 km toward the Earth's equator. However, the precision of our measurements of the variation in total solar luminosity (the so-called solar constant) over a whole solar cycle is still only about 1 percent; practically nothing is known about variations in the solar constant that might accompany fluctuations in solar activity occurring on time scales of 10^8 years.

The precise measurement of the solar irradiance by an instrument on the Solar Maximum Mission (SMM) has shown that luminosity variations in the range from 0.01-0.1 percent occur over periods of days and are associated with the growth and decay of large sunspot regions. But these results extend only over 1 year of the 22-year period of just one cycle. Obviously, the systematic and detailed scientific study of the solar cycle has just begun.

The second aspect of the investigation of the activity cycle is the study of the variation of the solar dynamo itself, which underlies the magnetic configuration of the convection zone and atmosphere. We discuss this important topic next.

4. The Solar Dynamo

Two factors combine with the convective motions present beneath the surface of the Sun to drive the solar magnetic dynamo. First, the Sun is rotating, and the interaction between the circulation in the convective zone and this overall rotation produces a differential rotation, in which the rotational period of the solar equatorial regions is shorter than that at higher latitudes. Second, the gas in the solar interior is highly electrically conducting--a property that "freezes" the magnetic field to a particular parcel of material as it moves about. Figure 1.6 indicates schematically how these mechanisms operate to cause the solar magnetic field to alternate between poloidal and toroidal configurations.

Theoretical studies of the Sun's hydromagnetic structure have shown that a variety of combinations of convection and nonuniform rotation within the spinning Sun can produce magnetic fields whose surface behavior mimics the migrations and reversals of the fields actually observed at the surface of the Sun. Further progress in the theory of stellar magnetic cycles now awaits the development of the dynamical theory of convection and circulation in the stratified envelope of a spinning star, to show which of the many "plausible" fluid motions used to demonstrate the generation of stellar fields is the motion

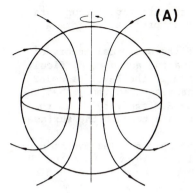 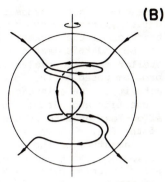

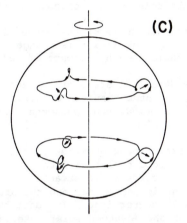

FIGURE 1.6 Differential rotation stretches an initially
poloidal solar magnetic field (A) to produce a toroidal field
(B); the twist of rising convective cells produces a small
poloidal field component (C).

that actually occurs. Progress in this direction depends on
continuing studies of the activity cycles in other stars to
determine the range of variation to be expected. More directly, progress depends on observing the large-scale circulation
at the surface of the Sun (expected to be some 1-100 m/sec) in
addition to the well-known nonuniform rotation of the Sun, so
that theorists will be able to check their dynamical models as
they are developed. An additional and equally important observational study is the splitting of the p and g modes of the
solar oscillations at the surface of the Sun, from which the
rate of change of angular velocity with depth may be deduced;
it is essential to know the depth variation of angular velocity

as a check on the dynamical theory, as this information enters directly into the dynamo equations that predict the form and behavior of the magnetic fields.

Beyond this immediate problem of the dynamical theory of stellar magnetic cycles is the fascinating and more difficult problem posed by the long-term changes in the level of solar activity. The existence of these changes would imply that the convection and circulation within the Sun can vary so as to allow less or more of the field to escape through the surface of the Sun.

The study of the mechanisms by which the solar dynamo operates, and how it varies with the solar cycle, can be begun by a Solar Interior Dynamics Mission. Three fundamental scientific questions should be addressed:

(a) How does the solar dynamo operate to produce a single magnetic cycle?

(b) How does the structure of the convection zone change over a solar cycle, and how does this change influence the operation of the dynamo and the outputs of radiation, plasma, and magnetic fields?

(c) How do the dynamical state of the convection zone and the operation of the dynamo change over periods of hundreds to thousands of years to bring about phenomena such as the Maunder Minimum?

In order to address these questions, increased theoretical activity, together with correlative data from many other sources (such as historical records of solar activity, observations of coronal and heliospheric phenomena, and the study of stellar activity cycles) will be required. In addition, observations must be continued on the Advanced Solar Observatory (ASO), after a solar Interior Dynamics Mission, in order to determine the characteristics of the solar dynamo over a full 22-year solar cycle.

5. Activity Cycles in Other Stars

What we now know about activity cycles in stars other than the Sun is due entirely to Olin Wilson's monumental program of synoptic monitoring of late-type main-sequence stars during the 11-year period 1966-1977. Wilson observed some 50 stars of spectral types F8-MO with narrow-band filters centered on the chromospheric H and K lines of Ca II. Despite the relatively short time baseline and inevitable data gaps, he observed cycles of 7-14 years' duration in a number of stars cooler than spectral type G2 (the spectral type of the Sun).

More recently, a program of daily monitoring of the calcium H and K lines in Sun-like stars has revealed variations arising from the growth and decay of active regions and their rotation onto or away from the visible disk of the star, thus providing a technique to measure stellar rotation rates (together with differential rotation rates, if a solar-type latitude dependence of starspot location on activity level also occurs in other stars).

As a result of these initial investigations, which verified the feasibility of stellar-activity studies, we believe that the following problems can be addressed in the 1980's and 1990's with well-conceived observing programs:

(a) Theoretical understanding of stellar dynamos will benefit greatly from a determination of how the period and relative amplitude of activity cycles depend empirically on such fundamental stellar parameters as luminosity, effective temperature, gravity, mean magnetic field, equatorial rotational velocity, age, and chemical composition.

(b) It is important to study the manifestations of stellar activity cycles at different levels of the atmosphere: chromospheric phenomena at optical wavelengths from the ground and in the ultraviolet with the International Ultraviolet Explorer, Space Telescope, and a far-ultraviolet spectrograph in space; coronal phenomena with future x-ray observatories, such as the Advanced X-Ray Astrophysics Facility; and phenomena at many atmospheric levels with Very Large Array (VLA) radio observations.

(c) As previously mentioned, nightly monitoring of the H and K line emission from stars can reveal individual active regions rotating on and off their visible disks as well as the growth and decay of active regions. It would be important to determine whether activity cycles can be simply described in terms of the changing fractional area of plages or whether plages and the chromospheric network themselves change qualitatively during activity cycles.

(d) It is difficult to study long-term changes in the solar activity cycle, but a study of a large sample of stars similar to the Sun might reveal equivalent information. For example, it might be feasible to determine the statistical probability and typical duration of Maunder Minimum absences of activity.

6. Solar Variability and Terrestrial Weather

Conditions on and near the Earth are determined first of all by the radiant energy from the Sun--sunlight--of some 0.14 W/cm^2. That radiative flux has heretofore been assumed to remain constant with time, except for a slow increase with the general aging and evolution of the Sun over its lifetime of about 10^{10} years on the main sequence.

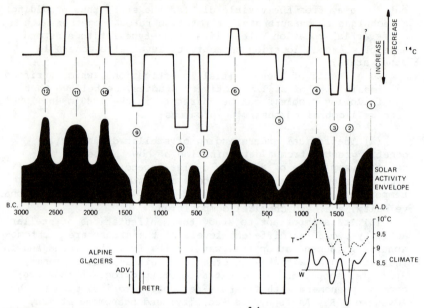

SOLAR ACTIVITY AND CLIMATE?

FIGURE 1.7 The rate of production of ^{14}C in the upper atmosphere by the bombardment of Galactic cosmic rays varies inversely with solar activity. The quantity of ^{14}C sequestered in three rings, when corrected for such factors as the changing dipole moment of the Earth and the radioactive decay of the isotope, yields a measure of solar activity over the past 5000 years. Top panel: The Maunder ② and Sporer ③ minimum established from optical observations of the Sun and auroral frequency provide a "modern" verification that the technique reveals real changes in solar activity. Because annual ^{14}C production is strongly buffered by the atmospheric and oceanic reservoirs, individual sunspot cycles do not appear. The center panel represents the same temporal variation of the ^{14}C proxy of solar activity smoothed out by about 50 years. The bottom panel shows mean European climate as measured by the advance and retreat of glaciers, by historical inferences of mean annual temperature, (T), and by the recorded severity of northern European winters (W).

The renewed interest in the connection between weather and climate, on the one hand, and the level of solar activity, on the other, is due to the striking correlation between climate and the general level of solar activity demonstrated by the analysis of geologic and botanical records, as illustrated by Figure 1.7. Although no causal relation between the weather and the solar luminosity (or any other solar emission that varies with the activity cycle) has yet been established, the strong correlation observed must cause us to question the assumption that the solar luminosity does not change over periods of hundreds of years. A synoptic study of the total luminosity of the Sun (from infrared to ultraviolet wavelengths) is of fundamental importance at this first level of the study of solar-terrestrial relations. The problem is so basic to the physics of stars and stellar activity and to the understanding of climate of the Earth that it should be given high priority on a long-term basis. As we have pointed out previously, a complementary program to measure the relative brightness of numbers of nearby main-sequence G stars is potentially a very powerful tool for studies of this type.

The more traditional aspects of solar-terrestrial relations have to do with the well-known large fluctuations in the output of the Sun at EUV and x-ray wavelengths and in the corpuscular emission from the Sun (the solar wind and the solar "cosmic rays"). These violent fluctuations affect the terrestrial magnetosphere, ionosphere, and upper stratosphere, producing intense heating during periods of enhanced solar activity. Indeed, there is statistical evidence that such effects may penetrate in subdued form into the lower atmosphere. Solar-wind fluctuations--particularly of magnetic-field direction--can induce substorms lasting a few hours that enhance energetic-particle deposition, ionospheric plasma production, and winds in the high-latitude thermosphere.

The Earth's atmosphere, ionosphere, and magnetosphere also respond to changes in large-scale solar-wind and magnetic-sector structures that rotate with the Sun. The polarity of the solar-wind magnetic field reverses abruptly across the magnetic neutral sheet that lies near the ecliptic plane. This neutral sheet has a number of folds determined by the way in which the solar-wind magnetic field happens to connect to the weak large-scale solar field. Each time the folded neutral sheet passes over the Earth, the Earth finds itself in a sector of reversed polarity. Since sector structure persists for several solar rotations, sector crossings tend to recur every 27 days (the solar rotation period), leading to a 27-day periodicity in various geomagnetic activity indices. Recently, it has also been proposed on the basis of statistical studies that the product of the vorticity and area of stratospheric wind systems increases after sector boundary crossings.

Variations in the Earth's magnetic field play a role on a still longer time scale. Every few hundred thousand years, the geomagnetic field changes polarity. When the field is very small, the Earth's magnetosphere has a radically different structure. Since the area accessible to energetic particle deposition and auroral activity is then much larger, the effects of solar activity noticed during the present epoch of a normally large geomagnetic field should be very pronounced during geomagnetic field reversals. Finally, on the longest time scales, variations in the solar wind over the age of the solar system must be considered. It was probably substantially more intense in the distant past, as suggested by observations of the dependence of the rotation rate upon age in other stars with convective outer layers and, presumably, stellar winds. Such studies may eventually further our understanding of the evolution of the magnetosphere and atmosphere of the Earth.

C. Nonthermal and Dynamical Phenomena in the Solar Atmosphere

The occurrence of a temperature reversal in the solar chromosphere, the million-degree temperatures characteristic of the corona, and the expansion of the corona into the solar wind all demonstrate that conditions in the solar atmosphere are far removed from local thermodynamic equilibrium (LTE). Nonthermal phenomena, which include the propagation and dissipation of waves, the presence of macroscopic flows (e.g., coronal current systems), and the confinement of the plasma by magnetic forces, have a major influence on the structure, dynamics, and energy balance of the atmosphere.

A series of observational programs, most notably the 1973 Skylab mission, have greatly expanded our understanding of nonthermal and dynamical phenomena in the solar atmosphere and have enormously sharpened the focus of an observational program for the 1980's. The central issues that must be addressed are summarized below:

• Because the structure of the chromosphere is dominated by the configuration of the magnetic field, which we now know to consist of basic elements of very small size (less than a few hundred kilometers), observations with very high spatial resolution are needed to guide and to test models of energy transport in the chromosphere.

• Newly emerging magnetic flux is accompanied by highly variable emission of EUV and soft x-ray radiation from small loops (with structures on a scale of 1000 km or less) called bright points. These structures, first detected by instruments on Skylab, may play a significant role in the magnetic evolution of the atmosphere and in mass transport into the solar wind but are too small to be fully resolved by present x-ray and EUV observations.

• Observations carried out by OSO-8 and SMM have shown that acoustic waves carry insufficient energy to heat x-ray-emitting coronal loops; in order to test alternative mechanisms involving the dissipation of energy in Alfvén waves or coronal current systems, we need improved models of the configuration of coronal fields and of the thermal structure of the coronal

loops themselves. These objectives require observations with
very high resolution (about 100 km) of the vector configuration
of the photospheric magnetic field at optical wavelengths and
of chromospheric and coronal structures at EUV and soft x-ray
wavelengths.

· Most current models of chromospheric and coronal
structure are static. We know, however, that mass transport
within the chromosphere, current systems in the corona, and
mass exchange between the chromosphere and corona and between
the corona and solar wind may play a dominant role in the
energy balance of the atmosphere. Observations with very high
spectral resolution for determination of spectral-line profiles
in small structures (100 to 500 km) are necessary to guide the
development of dynamical theories.

· It is generally accepted that the energy released from
flare processes is stored in force-free magnetic fields in the
corona and impulsively transformed into the kinetic energy of
nonthermal particles. (The relative importance of electrons
and nuclei is unclear, but electrons are thought to predomi-
nate.) The currents associated with the preflare configuration, and the reconnection process associated with particle
acceleration, evidently occur on a very small scale (about 100
km) and will require observations with very high spatial and
temporal resolution at soft x-ray and hard x-ray energies and
at radio frequencies to guide and test theory. In order to
elucidate the role of the acceleration of atomic nuclei in
flares, gamma-ray observations of high spectral resolution are
also required.

· While the general features of the large-scale structure
of the corona (such as coronal loops and coronal holes) are now
understood, the detailed dynamical behavior of the corona (such
as the mechanisms responsible for the acceleration and composition of the solar wind and the cause and role of coronal
transients) is not. To address these questions, a combination
of in situ observations as close to the Sun as possible,
together with high-resolution observations of the global
configuration of the corona, are required.

The common element in this brief review of nonthermal and
dynamical phenomena in the atmosphere is the need for higher
resolution observations (a) to determine the fine structure of
the chromospheric and coronal magnetic fields, (b) to determine
the dynamical behavior and fine structure of chromospheric and
coronal material, and (c) to determine the nature of the reconnection and particle-acceleration processes operating in flares.
It is clear that the necessary observations must cover a wide
range of energies, from the radio to the gamma-ray spectral
region, and must be made simultaneously because of the dynamical
nature of the phenomena to be studied. These requirements can
be largely met by two kinds of observational programs: (a)

high-resolution observations of solar magnetic structure with
ground-based optical magnetographs and with the VLA and (b) a
long-duration ASO in space, capable of the very highest
resolution attainable from Earth orbit (ultimately approaching
0.1 arcsec--70 km on the Sun--at most wavelengths) over the
full spectral range from infrared wavelengths to gamma-rays,
operating in concert with the VLA as appropriate, to study
nonthermal and dynamical phenomena in the solar atmosphere.
However, even instruments on the ASO will not be able fully to
resolve some important solar phenomena; a unique opportunity to
attain an order-of-magnitude improvement in resolution (about 7
km--equivalent to 0.01 arcsec from Earth orbit) and to make in
situ measurements directly in the corona is presented by the
encounter of the Star Probe mission, which will approach within
3 solar radii of the photosphere.

Finally, the study of chromospheres, coronas, and flares in
other stars having a variety of such fundamental parameters as
age, temperature, mass, composition, surface gravity, and
internal configurations should help elucidate how coronas and
chromospheres are heated and evolve.

1. Chromospheric Structure and Dynamics

Although the photosphere and chromosphere are layers of almost
negligible geometrical thickness compared to the radius of the
Sun (often being compared to the outer skin of an onion), they
are of fundamental importance to the energy, mass transport,
and heating of the entire outer solar atmosphere, including the
solar wind. The energy and mass flux that ultimately produce
the corona and solar wind move through these lower atmospheric
regions, leaving their signatures there; the energy dissipated
in the photosphere and chromosphere is roughly 10 times greater
than that required to heat the entire corona and solar wind.

The structure of the photosphere is dominated by the
granulation and supergranulation cells already discussed in
connection with convective processes. The magnetic field is
strongest in the cell boundaries (which presumably represent
upflowing material), and the "bright chromospheric network"
overlies these regions of strong field. The network is visible
in lines emitted by material ranging in temperature from the
low chromosphere (about 6000 K) up to 5×10^5 K, indicating
that the configuration of the magnetic field already dominates
the structure of the atmosphere in the low chromosphere and
that spherically symmetric models are not appropriate for any
portion of the atmosphere.

2. Energy and Mass Transport in the Photosphere and
Chromosphere

Although acoustic waves have been ruled out as the principal
agent responsible for heating the corona, they may still play
an important role in the chromosphere. The photospheric
granulation seems to describe a turbulent velocity field;
theory predicts a spectrum of sound waves that are in a frequency range that produces roughly the right acoustic dissipation as a function of height in the lower atmosphere. That is,
computed dissipation and computed net radiative losses are approximately equal, according to the best available atmospheric
models. Furthermore, wave propagation is now thought to be
responsible for the magnitude and height dependence in the
chromosphere of the "microturbulent" velocity, calculated by
comparing observed linewidths with computed widths and attribut-
ing the difference to a nonthermal velocity field. Determination of the spectral characteristics of acoustic waves in the
chromosphere will require determination of the granulation
spectrum at high wavenumbers and frequencies, which will in
turn require spatial resolution in the subarcsecond range.
Knowledge of the high-frequency components of the granulation
spectrum is required for accurate estimates of the energy flux
in sound waves generated by this turbulence. In addition,
observations of photospheric and chromospheric lines are needed
at subarcsecond resolution with high sensitivity to obtain
reliable power spectra yielding unambiguous evidence for the
propagation of waves in these regions. A proper study of the
propagation and dissipation of waves in the photosphere and
chromosphere requires a self-consistent treatment of non-LTE
radiative transfer; this task may require the next generation
of computers. Finally, the effect of a magnetic field on wave
propagation in a realistic model chromosphere must be determined.
    A second aspect of chromospheric heating is the shock dis-
sipation of acoustic waves. At present, only indirect evidence
for shock dissipation is available (e.g., the apparent presence
of sound waves with the "right" dissipation length, a small
density scale height). From this evidence and supporting
theoretical calculations, many (though not all) solar physicists now feel that the shock dissipation of initially sinu-
soidal acoustic waves generated by photospheric turbulence
provides sufficient energy to produce the initial chromospheric
temperature rise, as well as to balance the considerable radiative dissipation in the photosphere. However, it appears that
the dissipation of acoustic waves cannot heat the entire upper
chromosphere, where the so-called Lyman-alpha temperature
plateau at about 2×10^4 K occurs, and furthermore may not be
able to account for the energy dissipated in the region where
the Lyman-alpha wings and the resonance lines of Ca II and

Mg II are formed at about 6.5×10^3 K. The high-resolution observations of the Solar Optical Telescope will provide theory with empirical guidance in assessing the role of magnetic heating (via current-sheet dissipation) and of energy transport from the hotter corona into the upper chromosphere.

It is clear that mass transport plays an important role in the energy balance of the chromosphere. Above the temperature minimum, much of the emission associated with the chromosphere comes from spicules--outwardly moving structures observed on the solar limb, which are thought to comprise the chromospheric network on the disk. When observed at the limb, the spicules typically rise at velocities of about 30 km/sec for times of 5 min or so, then seem to disappear or even occasionally fall back. However, the net flux of unresolved spicules is outward, so much so that one of the long-standing and unsolved problems of solar physics is that the integrated efflux of mass estimated from spicules is an order of magnitude larger than the efflux of mass from the Sun in the solar wind measured at the Earth's orbit. Where does the "upward" flowing mass go? Recent observations show rapidly rising gas jets (superspicules) with velocities up to 400 km/sec and, more numerous and in different locations, falling material with smaller but still impressive velocities up to 100 km/sec. These speeds are highly supersonic in the upper chromosphere. Future research should focus on at least two questions concerning spicules and these gas jets:

(a) What characterizes the flows in the chromosphere that preserve mass balance in the corona, while providing for the leakage due to the solar wind?

(b) What physical processes cause the rising and surge-like behavior of spicules and gas jets?

A number of other problems of chromospheric structure and dynamics, which we cannot discuss in depth here, will be addressed by the observations of the Solar Shuttle Observatory and the ASO; examples include the following:

(a) The interpretation of the radiation from ionized helium, to determine the helium abundance and the ionization structure of the upper chromosphere;

(b) Development of an improved model of the minimum-temperature region in the lower solar chromosphere; and

(c) Development of a full non-LTE model of radiation transport in the chromosphere that includes the effects of atmospheric fine structure, differential velocity fields, and multidimensional radiative transfer.

3. Coronal Structure and Dynamics

In the past decade, observations made with rocket and satellite instruments operating at UV, EUV, and x-ray wavelengths have demonstrated more clearly the fundamental role that magnetic fields play in defining the structure and the mass and energy flow in the chromosphere and corona, resulting in the identification of coronal holes as the source of high-speed solar-wind streams. Differences in the physical conditions in the chromospheric and coronal layers over different regions of the solar surface appear to be intimately related to the configuration and strength of the magnetic fields in these areas. The corona is divided into three classes of regions: (a) active regions and coronal bright points, characterized by strong coronal magnetic fields with closed configurations (coronal loops); (b) quiet regions, which have weak coronal magnetic fields that appear to be closed on a large scale; and (c) coronal holes, which are associated with weak coronal magnetic fields having a diverging open configuration. Regions with strong fields are characterized by enhanced chromospheric and coronal radiative output, moderately high coronal densities, and characteristic coronal temperatures of 2.5×10^6 K. Quiet regions are characterized by average chromospheric radiative output, intermediate coronal radiative output and densities, and characteristic coronal temperatures of 1.5×10^6 K. Coronal holes are characterized by average chromospheric radiative output, low coronal radiative output and densities, and characteristic coronal temperatures of 1×10^6 K. The differences in the physical conditions in these regions must be due in part to the ability of magnetic fields to channel the flow of mass and energy in the corona and possibly to the direct dissipation of magnetic energy stored in the field.

High-resolution observations of active regions made in the past decade have demonstrated that the coronal layers of active regions (within 1 to 2 solar radii of the surface) consist of elemental structures in the form of magnetic flux tubes or loops in closed configurations (i.e., with both footpoints rooted in the chromosphere). One of the surprises of the 1970's was the discovery of cool loops, filled with plasma at temperatures below 10^6 K, which appear to be dynamically and thermally unstable, with material flowing downward into the chromosphere. Sunspots, in particular, are often located at one footpoint of these cool loops.

Empirical and theoretical studies of these loops have demonstrated that the dynamics of the plasma contained in these loop structures is fundamental to their nature and that future observational and theoretical programs must take this factor into account. The exchange of mass and energy between the chromospheric and coronal levels in these structures appears to play a major role in determining the temperature-density structure of the coronal layers of active regions.

Among the more intriguing discoveries made by high-resolution rocket and satellite experiments in the past decade are the coronal bright points, small regions of intense x-ray and EUV emission distributed more or less uniformly over the solar disk. These regions, with characteristic dimensions of 20,000 km (30 arcsec), appear to be miniature active regions with lifetimes measured in hours; they often exhibit flarelike activity in which the EUV and x-ray brightness increases by one or more orders of magnitude over a period of a few minutes. Studies of the rate of emergence of bright points from observations made at different times in the last decade suggest that the number of bright points on the Sun is anticorrelated with the sunspot number and that the magnetic flux associated with these features is a major contributor to the total amount of magnetic flux being brought to the surface throughout the solar sunspot cycle.

Another important feature of bright points is their association with polar plumes, which contribute a significant fraction of the mass in polar coronal holes. It appears that bright points may provide an important contribution to the outward mass flux that is observed in high-speed solar-wind streams originating in coronal holes. If so, investigations into the physics of the bright points located in coronal holes may have a significant impact on our understanding of the physical processes by which the solar wind is generated there.

Although it is now known that high-speed solar-wind streams originate in coronal holes (regions of rapidly diverging open magnetic field configurations), the source of the low-speed component of the solar wind, the mechanisms responsible for solar-wind acceleration, and the large variations in the chemical composition of the wind are not understood.

The study of the chemical compositions of the chromosphere, corona, and solar wind can provide insights into the dynamical processes occurring in the atmosphere. Although theoretical diffusion models have predicted composition differences in the chromosphere-corona transition region and in the corona, no substantial evidence for such differences has yet been found. This suggests that sufficient mixing of the atmosphere takes place because of dynamical processes that lead to the interchange of mass between the chromosphere and lower corona, so that differences in chemical compositon between these layers are eliminated.

The delineation of the acceleration and escape mechanisms for solar mass loss and their relationship to the mean solar abundance distribution in the solar wind and in solar cosmic rays are important new areas of solar research.

The chemical composition of the solar wind is of interest in the broader context of solar abundances and enrichment of the interstellar medium. The solar-wind observations made to date have shown that the abundance of helium in the solar wind is 30 to 50 percent smaller than the normal solar abundance and

that it varies widely on a variety of time scales. The largest
variations (as much as an order of magnitude) occur on a scale
of hours and appear to be related to solar-flare activity,
while variations by nearly a factor of 2 may occur on the time
scale of the solar cycle. Although observations of other
isotopes and elements (^{3}He, Ne, O, Ar, and Fe) have been made
only occasionally, similar degrees of fluctuation are clearly
present.

4. Energy and Mass Transport in the Transition Region and Corona

The source of coronal heating provides a classic example of a
central problem in astrophysics whose solution has proven to be
elusive but whose pursuit has been extraordinarily productive.
Current models of coronal heating emphasize mechanisms that
involve the release of energy stored in the coronal magnetic
field either by the dissipation of Alfvén waves, by the dissipation of field energy represented by coronal currents, or in
some type of slow reconnection process.

For example, a comparison of calculated coronal-field
configurations (based on measurements of photospheric fields)
with coronal loop structures implies that coronal fields are
force-free and that large currents must flow in the corona.
The dissipation of these currents by classical ohmic heating
cannot provide sufficient energy to heat a coronal loop;
however, turbulent resistivity can give rise to significant
heating.

Active-region loops may exist in a quasi-steady state in
which new current generated by twisting of magnetic fields
through photospheric motions is dissipated by turbulent resistivity in the corona, providing the energy necessary to balance
the energy lost through conduction and radiation. Similar
mechanisms may be at work in x-ray bright points, which differ
from active regions (and perhaps all large-scale closed coronal
structures) only in that they are compact and probably confined
by highly twisted fields.

Field reconnection provides a somewhat different mechanism
for magnetic dissipation. If the twist of a magnetic flux tube
exceeds a modest amount, it becomes unstable and rapidly forms
enhanced current sheets in which oppositely directed field elements are brought together. This process thus leads to rapid
field reconnection and energy release across the neutral sheet.
Such behavior could be the cause of bright-point flares, which
are observed to exhibit many characteristics of active-region
flares. If such a process occurs continuously, it could also
produce the quasi-static heating required to sustain bright
points and coronal loops.

Both current dissipation and magnetic reconnection may play a role in the conversion of magnetic energy into heat in the corona, the former producing quasi-stable heating and the latter being associated with more dynamic field instabilities on all scales, from bright points up to major flares.

High-resolution observations acquired over the past decade have led to renewed interest in problems of mass transport in the chromosphere and corona and have shown that mass transport processes play a critical role in controlling the physical state of the outer solar atmosphere. It is now clear that the outward flow of the solar wind represents a major coronal energy loss from open-field regions such as coronal holes and is directly related to the low coronal temperatures and densities found in the inner coronal layers of these regions. High-resolution observations of active and quiet regions have shown that there is a continual interchange of mass and energy between the chromosphere and corona, often occurring in fine-scale (subarc-second) structures associated with regions of strong magnetic fields or somewhat larger (few-arcsecond) structures in active region loops and sunspots. The role of dynamic phenomena such as spicules, "superspicules," and surges in coronal mass deposition is unknown. There is strong circumstantial evidence that chromospheric evaporation driven by thermal conduction plays a primary role in determining the temperature-density structure of coronal loops in quiet regions, active regions, and bright points, but direct empirical confirmation of this process needs to be obtained. In general, when the chromosphere and corona are observed with high spatial resolution (approximately 1 arcsec), it is found that dynamical motions and mass flows appear to be critical elements in the physical state of the plasma. What appears as microturbulence or nonthermal broadening of line profiles in measurements made with low spatial resolution seems to be caused by macroturbulence and mass flows in observations of higher resolution. What was once judged to be a quasi-static atmosphere in hydrostatic equilibrium is gradually coming to be recognized as a dynamic atmosphere, with continual interchange of mass and energy between the chromosphere and corona.

5. Chromospheres and Coronas in Other Stars

The study of stellar chromospheres and coronas has been a particularly exciting and rapidly advancing subject, and it has produced surprises with far-reaching consequences. The existence of chromospheres in a wide range of stars was clearly demonstrated before 1978 by ground-based observations of the H and K emission lines of Ca II and by observations of the Mg II resonance lines in emission with OAO-2, Copernicus, and balloon experiments. These data implied that chromospheres exist in

most or all stars cooler than about spectral type F5 (7500 K),
although some stars as hot as spectral type FO show evidence
for chromospheres.

The International Ultraviolet Explorer (IUE) satellite,
launched in January 1978, has greatly expanded our knowledge of
the outer atmospheric layers of stars through observations of
emission lines formed at temperatures of 4,000-250,000 K. So
far, IUE observations indicate that only dwarf stars cooler
than spectral type FO, together with F- and G-type giants and
some supergiants, show emission lines characteristic of transition regions (and presumably coronas, if the solar analogy is
valid). The apparent absence of transition regions (and presumably also coronas) in K-M giants and in most supergiants may
indicate that winds, rather than radiation from hot coronas,
are the dominant cooling mechanism to balance nonradiative
heating.

Early x-ray experiments on rockets and satellites
(Netherlands Astronomy Satellite, Small Astronomical
Satellite-3) detected coronal emission from Capella, several
dMe stars during flares, and, unexpectedly, from Vega and
Sirius. The most exciting result from HEAO-1 was the discovery
that the RS CVn-type binary systems, in which one or both
components are slightly evolved off the main sequence and
similar to or slightly cooler than the Sun, are powerful x-ray
emitters with coronal emission measures up to 10^5 times that
of the quiet solar corona.

Results from the Einstein x-ray observatory (HEAO-2) have
produced major surprises. In addition to stars of types
previously known to have coronas, Einstein observations revealed
coronas in young O stars, Of stars, Wolf-Rayet stars, and B-type
supergiants. These data show that coronas exist in a much wider
range of stars than previously thought and suggest that coronas
may be heated by different mechanisms in different regions of
the H-R diagram.

During the next decade, observational programs carried out
with Space Telescope, the Extreme Ultraviolet Explorer, a
far-ultraviolet spectrograph in space, and an Advanced X-Ray
Astrophysics Facility will provide a much more coherent picture
of coronal emission and stellar winds in stars over the entire
H-R diagram. Comparative studies of parameters associated with
the solar corona and wind and those associated with stellar
coronas and winds should provide critical tests for models of
coronal heating.

6. The Flare Phenomenon

The solar flare represents the most conspicuous aspect of the
activity of the Sun and is the central and final link in the
shedding of the solar magnetic field, which leads to the decay

of the Sun's activity cycle. Flares are significant, not only for their geophysical effects and the many secondary solar phenomena that they generate but also for the opportunity that they offer to study processes important to astrophysics and plasma physics generally.

Despite its complexity, a conceptual flare model (based on high-resolution observations from the ground and from the Orbiting Solar Observatory, Skylab, and SMM spacecraft) has recently gained wide acceptance. The basic elements of this model are as follows: prior to flare onset, energy is stored in a current-carrying magnetic field that is in a metastable state. The sudden reconnection of this field releases its free energy, which appears in the form of energetic particles (mainly electrons, according to most models). These particles subsequently interact with the atmosphere to produce in situ heating and bursts of microwave radiation and hard x-rays. The energy flux in this beam is high and leads to explosive evaporation of the chromosphere, producing a dense, hot (10^7 K or more) plasma. This primary energy deposition in turn accounts for many subsequently observed flare phenomena, including soft x-rays, chromospheric radiation, far-ranging coronal mass transients and radio emission, and solar cosmic-ray acceleration. Finally, the coronal region surrounding the initial event is left filled with a magnetically confined hot plasma, which maintains the decay phase of the flare for hours.

Important questions do remain, of course. Some observations suggest that there is a preheating phase before the explosive energy release, as revealed by brightening of coronal loops, but the reality of this pheomenon has not yet been firmly established. In addition, the relationship of flare initiation to apparently preliminary coronal transients has not been made clear. Both of these possible precursors seem to be evidence of a very-large-scale agitation of the active-region magnetic field before a flare. The critical physical processes occur on a small scale (100 km or less) and, in some cases, are observable only at radio, far-ultraviolet, or x-ray wavelengths. Detailed diagnostic study of these critical phenomena through coordination of the high-resolution observations is now necessary to determine the precise physical nature of each phenomenon.

Magnetic reconnection is the central process (both in the laboratory and in astrophysics) for the catastrophic release of the energy stored in a magnetically confined or stressed plasma; in our discussion of basic physical processes, it therefore deserves a conspicuous place and some explanation. Under a wide variety of conditions, plasma tends to be frozen to magnetic field lines. This constraint is, however, not absolute; if it were, then perfect plasma confinement would be possible in the laboratory, the Earth's magnetosphere would be isolated from the solar wind, matter would be less likely to escape from evolving stellar systems, and convection would not occur in magnetized stars. Quite generally, the constraint breaks down locally in the vicinity of singular layers (sometimes called neutral sheets) where the magnetic field is locally orthogonal to an (unstable) eigenperturbation of the global plasma structure. In the vicinity of these singular regions, the field is freed from the plasma and its inertia and can rapidly reconnect into a topology that permits the spontaneous relaxation of stresses. This reconnection, also called magnetic tearing, may occur either gradually or explosively. When it occurs explosively, it can lead to auroral substorms and solar flares or to disruptions of the discharge channel in laboratory experiments with tokamaks; when it occurs gradually in interstellar space, stellar convective layers, or stellar surfaces, the results include the quasi-static but greatly enhanced dissipation of magnetic energy and the rearrangement of the field topology. The primary direct energy output of dynamic magnetic reconnection is accelerated electrons. These electrons, through collisional and collective-mode heating and bremsstrahlung, provide the observed low-energy and hard-radiation signatures of the classical solar flare.

A fairly clear morphological picture (and a partially resolved quantitative description) of the preflare state and of the quasi-thermal decay phase of a flare have been obtained from extensive analyses of soft x-ray and EUV observations. However, coordinated full-spectrum diagnostic investigations have been rare, and the present observations do not have sufficient spatial discrimination to resolve the thermal structure of coronal loops. The magnetic geometry has been documented in a number of cases, and the discovery of single-loop flares has clarified the instability potential of an elemental sheared field (opening up new generic connections with toroidal-pinch laboratory experiments). The density increase resulting from a flare has shown the importance of the chromospheric evaporation response to the primary energy release. The rough temporal and spatial variability of flare hard x rays has been determined, indicating that a major impulsive-phase output is a flux (about 10^{29} ergs/sec in a major flare) of 1-100 keV electrons. Such electrons, together with the plasma waves (leading to Type-III radio bursts) they generate in traveling through the corona, have been observed in interplanetary space. During the last few years, gamma-ray detectors and radio interferometers have begun to produce data on the initial flare output, demonstrating the existence of MeV nucleons and very localized 100 keV gyrating-electron populations.

A major discovery of the past decade was that coronal transients are a much more frequent phenomenon than previously realized and that these events are often associated with a restructuring of the outer corona. Coronal transients are mass ejections seen in the outer corona with velocities that typically lie between a few hundred and a thousand kilometers/second. These ejections are generally associated with eruptive prominences, flares, or other types of impulsive coronal events. Because of their large energies (typically $10^{29}-10^{30}$ ergs, comparable with the energy liberated in other manifestations of flares), and frequent association with flares, it is apparent that coronal transients play an important role in the flare process. The sizes and trajectories of flare-associated transients indicate that large volumes of the corona above the flare site are affected by passage of a transient. These events are also a significant source of solar mass loss.

The impressive observational progress of the last decade has permitted a new level of physical inquiry into the dynamics of the solar atmosphere. In a sense, the quality of the measurements has brought solar physics into an era in which detailed diagnostic measurements (on the plasma-physics model) may be made and detailed theoretical comparisons attempted. This is not to say that new phenomena will not be discovered, but we are clearly at the threshold of an integrated investigation--uniting theory and experiment--of energetic solar-physics processes.

We have already described the kinds of observations necessary to further refine our knowledge of preflare and postflare energy deposition and decay processes. We do wish, however, to emphasize again the importance of the high-energy instrumentation for the observation of hard x rays and gamma rays that must be incorporated into the ASO before the next solar maximum and of collaborative observations between the ASO and the VLA.

7. Flares in Other Stars

Transient x-ray and XUV emission during flares has now been detected from several dMe flare stars, including UV Ceti, YZ CMi, Proxima Centauri, and from several RS C Vn systems. These flares have temperatures, x-ray light curves, and radio emissions similar to those of solar flares, but x-ray luminosities and coronal emission measures up to four orders of magnitude larger than is typical for large solar flares. For some flares, the optical emission is much greater than the x-ray emission, indicating that, as in solar flares, the stellar flares are cooled mainly by conduction to the chromosphere. However, some observed flares in the dMe stars have much more intense x-ray emission than optical emission, indicative of direct radiative cooling. Thus, the large geometric scale or high densities implied by the large emission measures in stellar flares may lead to cooling-phase behavior different from that of typical solar flares.

During the next decade, the study of flares in dMe and other stars in the context of solar-flare concepts and models should incorporate a number of research thrusts. Coordinated observing programs involving simultaneous gamma-ray, x-ray, UV, optical, and radio observations are needed to determine what types of stars (in addition to the dMe stars and the RS CVn systems) actually do flare and whether the physical properties of solar and stellar flares are similar. High-resolution spectroscopy in the x-ray and XUV regions of the spectrum are needed for accurate determinations of temperatures, densities, volumes, and other flare properties. With such data, one can begin to learn whether the flare mechanisms are similar to or different from those operating on the Sun. In particular, it is critical to know why flares in dMe stars and RS CVn systems are orders of magnitude more energetic than solar flares. Finally, we would like to know the extent of nonthermal processes in stellar flares, and why the ratio of x-ray to optical emission is so highly variable.

D. The Dynamics of the Heliosphere

The interplanetary medium consists of four principal components: (1) a thermal plasma (the solar wind), which is an extension of the solar corona flowing supersonically away from the Sun; (2) a magnetic field, which is that part of the solar magnetic field carried into interplanetary space by the solar wind; (3) energetic particles, which have been accelerated to high energies in the solar atmosphere (solar cosmic rays), beyond interplanetary space (Galactic cosmic rays), within interplanetary space, or within planetary magnetospheres; and (4) a neutral gas, which flows into interplanetary space from the interstellar medium. It is clear that the interplanetary medium is appropriately viewed as the outermost solar atmosphere (sometimes referred to as the heliosphere) and that its study represents an integral part of solar physics.

The key element in our improved understanding of heliospheric structure and dynamics and its relationship to coronal structure and dynamics has been a collaborative program of coronal observations—most notably on Skylab—and of in situ plasma, magnetic-field, and energetic-particle observations in interplanetary space. During the past decade, the region observed has been extended to within the orbit of Mercury and beyond the orbit of Saturn, but always in or near the ecliptic plane.

During the next decade this process of exploration should be greatly accelerated through the first in situ observations out of the ecliptic plane by the International Solar Polar Mission (ISPM) and through intrumentation on a Solar Coronal Explorer (SCE), which will, for the first time, provide global measurements of the solar-wind flow. In the early 1990's, this program of exploration may be extended directly into the corona itself, where the solar wind is accelerated, by a close-encounter space mission--the Star Probe--and by high-resolution observations of coronal transients, solar wind flow, and coronal magnetic structure with an ASO.

1. The Solar Wind and the Interplanetary Magnetic Field

In the past decade, two of the most significant advances in our understanding of the solar wind have resulted from studies of the spatial and temporal evolution of large-scale solar-wind structures (e.g., high-speed streams, interplanetary magnetic sectors, flare-produced interplanetary shocks) and from studies of the relationship of coronal holes to high-speed solar-wind streams and interplanetary magnetic sectors. The former studies have given us a good understanding of the basic physical processes important for the evolution of large-scale structures in the ecliptic plane and have developed a foundation for three-dimensional magnetohydrodynamic modeling of the evolution of such large-scale structures throughout the heliosphere. The latter studies have served both to resolve old problems and to present new ones.

Coronal holes have now been identified unambiguously as the long-sought solar sources of recurrent high-speed solar-wind streams, and this identification has led to new inferences (based on coronal observations) of the three-dimensional structure of the solar wind and interplanetary magnetic field, at least during the years surrounding the minimum in the solar-activity cycle. However, this identification has also led to the conclusion that the nonradiative energy flux (entering the corona from lower atmospheric layers) required to supply the energy carried away by the solar wind is nearly an order of magnitude larger (for high-speed streams from coronal holes) than has previously been assumed; hence, the solar wind dominates energy balance in coronal holes. This conclusion, of course, adds a new dimension to the classical coronal heating problem. The problem is further complicated by the rapidly diverging magnetic-field geometry (and flow geometry) that must be considered in discussions of the coronal expansion in coronal holes.

Clearly, one of the problems ripe for attack during the next decade is the determination of the three-dimensional structure of the heliosphere and the relationship of that structure to the large-scale structure, energy balance, and dynamical behavior of the corona. On the basis of recent, complementary observations of coronal holes and the solar wind, it appears that, near the time of solar-activity minimum, the interplanetary medium may be well ordered on a large scale, with the interplanetary magnetic field exhibiting predominantly one polarity in the northern hemisphere and the opposite polarity in the southern hemisphere and the solar-wind speed increasing with latitude from generally moderate values near the ecliptic plane to broad, relatively uniform high-speed streams at high solar latitudes. By contrast, it seems that, nearer solar maximum, this ordering of the interplanetary medium on the largest scale disappears together with the corresponding ordering of brightness structure in the solar corona (large coronal holes at high latitudes and bright, magnetically closed regions at low latitudes). The observational programs to be carried out by the ISPM, the Interplanetary Laboratory (IPL) of OPEN, and the SCE are indispensible to this scientific program.

Toward the end of the 1980's and in the early 1990's, it should be possible to initiate an intensive program to address a second major problem relating to the heliosphere--the location of the region of, and the delineation of the mechanisms responsible for, the acceleration of the solar wind, as well as the explanations of related phenomena, such as the large variations observed in the solar-wind composition. In order to address these problems properly, the direct, in situ observations of the Star Probe, together with the high spatial and spectral resolution and high sensitivity of the ASO, will be required.

2. The Propagation of Energetic Particles in the Heliosphere

The existence of large-scale ordering of the interplanetary magnetic field has important implications for the interplanetary modulation of Galactic cosmic rays. In the presence of a well-ordered magnetic field, gradient and curvature drifts (whose effects are not generally included in modulation theory) could be significant. This entire problem--the three-dimensional structure of the solar wind and the interplanetary magnetic field, and its influence on the modulation of Galactic cosmic rays--can be addressed in the next decade by ISPM and by the IPL of OPEN, which will provide in situ observations of the solar-wind plasma, the interplanetary magnetic field, and Galactic cosmic rays both in and out of the ecliptic plane. These direct observations need to be supplemented by observational and theoretical studies of the interplanetary acceleration of energetic particles (which modifies the cosmic-ray energy spectrum) and of the diffusion and scattering of cosmic rays through interaction with interplanetary fluctuations. This could be accomplished in part by an Advanced Interplanetary Explorer satellite, which could study the propagation of solar and Galactic cosmic rays in the interplanetary medium. In particular, it is important to determine whether the anomalous component of the cosmic-ray spectrum is accelerated in interplanetary space (from singly ionized atoms originating in the interstellar neutral gas); if not, the basic theory of interplanetary modulation of Galactic cosmic rays may face fundamental problems.

3. The Neutral Component of the Interplanetary Medium

In conjunction with studies of the interplanetary medium in three dimensions, it is important to emphasize studies of the penetration of the neutral interstellar gas into the inter-planetary medium. Observations with spacecraft UV photometers, together with steady-state theoretical models, have brought us to a good understanding of the basic physical processes involved in this aspect of the interaction of the heliosphere with the interstellar medium. However, it will be necessary in the next several years to develop time-dependent theoretical models (because of the several-year residence time of an interstellar hydrogen atom in the interaction region of the heliosphere) and to obtain higher resolution UV sky-background maps, in order to improve the accuracy of the deduced values of local interstellar gas parameters and to obtain the three-dimensional struc-ture of the solar-wind mass flux from such maps. The Extreme Ultraviolet Explorer, Space Telescope, and a far-ultraviolet spectrograph in space should all contribute important observational data relevant to this question.

4. The Loss of Solar Angular Momentum

Another problem that should be addressed in the next decade is the question of solar spindown, or the transport of angular momentum by the solar wind, which cannot at present be pursued without further observations to test current theories. Such observations can be carried out in and out of the ecliptic plane by plasma instruments on properly placed interplanetary probes, such as ISPM and the IPL of OPEN. In addition, if the white-light coronagraph on ISPM can measure the curvature of coronal structures out to some 20 solar radii, current theories can be put to a solid test. The variation of the angular-momentum transport over a significant fraction of the Sun's lifetime can be addressed indirectly by studies of lunar material and their record of the ancient solar wind.

VI. PROGRAMMATIC OPPORTUNITIES FOR THE 1980'S

The preceding section has presented the scientific goals that the Working Group believes should be pursued during the

1980's. Because the scientific programs required to pursue these goals depend so heavily on coordinated observations using a variety of techniques, a discussion of programmatic opportunities in solar physics for the 1980's must necessarily emphasize the coherence of these programs. We wish also to emphasize the importance to solar physics of ongoing programs at NSF and NASA; of approved programs, such as the Solar Optical Telescope (SOT) and the International Solar Polar Mission (ISPM); and of related programs, such as Star Probe and OPEN, which are normally reviewed by established committees of the National Research Council's Space Science Board.

We first present a summary of the major theoretical activities and observational programs that we believe will be needed during the coming decade and then discuss each program in some detail.

A. Summary

1. Theory and Modeling, and Laboratory Astrophysics

Theory must play an increasingly central role in the planned development of solar physics. Moreover, theory and quantitative modeling should guide the entire information chain--data acquisition, reduction, dissemination, correlation, storage, and retrieval--in order to assure the availability of coordinated data from diverse origins. The carefully planned Apollo Telescope Mount (ATM) observational and data-analysis programs, which made extensive use of workshops, furnish an excellent model. In this connection, we support the full exploitation of Solar Maximum Mission (SMM) observations through an extended program of data analysis and associated theoretical activity that includes broad community participation.

The Colgate Committee of the National Research Council's Space Science Board (Space Plasma Physics: The Study of Solar-System Plasmas, Volume 1, Reports of the Study Committee and Advocacy Panels, National Academy of Sciences, Washington, D.C., 1978) recommended that the level of theoretical effort in space plasma physics be increased. This recommendation has already been put into effect by NASA for solar-terrestrial physics; we applaud this initiative, urge that it be continued, and hope that it will be followed by a similar NSF initiative.

We also endorse expanded support for laboratory measurements of those atomic, molecular, and nuclear properties that are relevant to astrophysics; without such data, the major instrumental programs discussed here cannot achieve their full potential.

2. Ground-Based Facilities

The role of space observations in solar physics should reach full maturity as an equal partner with ground-based techniques during the 1980's; consequently, the majority of the major new initiatives that we have identified involve space observations. Nevertheless, ground-based solar astronomy remains scientifically vital and technologically innovative. Collaborative observational programs between spacecraft instruments and ground-based instruments such as the Very Large Array (VLA) and vector magnetographs, together with innovative ground-based programs such as the study of the solar oscillations, neutrino observations, and the study of stellar chromospheres and activity cycles, are central to the scientific strategy that we have developed. We summarize the major new ground-based initiatives that we have identified below:

(a) We strongly support the ^{71}Ga solar-neutrino experiment and urge that it be implemented as soon as possible. We further urge that ^{7}Li and ^{115}In experiments be implemented during the coming decade.

(b) We strongly support a vigorous program of solar research at the National Astronomy Centers. The Centers at which solar research is carried out--Sacramento Peak Observatory (SPO), Kitt Peak National Observatory (KPNO), and the High Altitude Observatory (HAO)--all enjoy strong and active support and participation by the solar-physics community. Although there are no staff members at NRAO who are currently active in solar research, a strong visitors' program of solar research utilizing the VLA has developed.

We also wish to stress the importance of university-operated solar radio and optical observatories and of independent observatories such as Mount Wilson to solar-physics research during the next decade. The research carried out at these observatories is innovative and frequently of a pioneering nature, and such institutions play a unique role in the training of young scientists.

We believe that the major development programs that we have identified can be accomplished within the continuing National Astronomy Centers and Grants Programs of the NSF; however, we are concerned that the present levels of support for instrumentation and detectors development and for technical support of observatory operations are inadequate in both the Centers and Grants Programs.

Among the major development programs for solar instrumentation discussed by the Working Group, we wish specifically to endorse the following, which are listed in order of priority:

(i) Ultra-high-resolution tachometers for observations of the velocity field of the Sun, with applications to studies of large-scale circulation, differential rotation, and the various solar oscillation modes.

(ii) Instrumentation that utilizes active optical elements and image-restoration techniques to obtain diffraction-limited resolution for studies of solar structure and motions.

(iii) One or more vector magnetographs for the study of the electric-current patterns associated with solar activity.

(iv) Improvements in the VLA to enhance its capabilities for solar observations.

(v) An extreme-limb photometer to study large-scale photospheric brightness variations and their influence on measurements of the shape of the Sun, as well as to study solar oscillations.

(vi) Development of a capability to make full spatially resolved polarization measurements at radio wavelengths at a suitable facility.

(vii) Instrumentation that can achieve improved sensitivity for the study of stellar magnetic fields.

(viii) The development of instrumentation for direct measurements of solar electric fields through observations of the Stark effect.

(c) Ground-based observations have been in the vanguard of the expanded interest of astronomers in stellar activity and stellar cycles; however, the unique requirements of long-term observational programs that this type of study imposes have made it difficult to accommodate such programs within the existing resources of the astronomical community. We believe that two measures should be taken to alleviate this problem:

(i) The construction of a dedicated stellar telescope of moderate aperture (about 2.5 m) for synoptic observations of stars, to be operated in conjunction with the solar programs of the National Astronomy Centers.

(ii) Observations with the Mount Wilson 1.5-m telescope have provided much of our present results on stellar chromospheres and stellar activity. The continued operation of this facility is now in doubt because of financial pressures on the private foundation that operates the Mount Wilson Observatory. We urge the relevant funding agencies to weigh carefully the various alternatives by which they could help assure the continued operation of this unique and important facility.

3. Space Facilities

(a) The Working Group concludes that the space program of highest priority for solar physics in the coming decade is the evolutionary development of an ensemble of Shuttleborne instruments in both the facility and Principal Investigator classes that may be operated together to form a Solar Shuttle Observatory, and that will be capable of simultaneous, high-resolution observations covering the electromagnetic spectrum from gamma-ray to near-infrared wavelengths. Once developed, this assembly should be placed on a free-flying satellite or space platform to form an Advanced Solar Observatory for the long-term study of the physics of solar and stellar activity.

(b) We endorse the recommendation of the SSB's Committee on Solar and Space Physics for a solar encounter and interplanetary cruise mission, the Star Probe, to penetrate as deeply as possible into the acceleration region of the solar wind. The major scientific objectives of the Star Probe of interest in the present context are the following:

• To obtain ultra-high-resolution (7-10 km) observations of the corona, chromosphere, and photosphere at solar encounter;

• To study the interplanetary medium and solar corona in situ, including the acceleration region of the solar wind;

• To obtain a direct measurement of the solar quadrupole moment; and

• To obtain extended stereoscopic observations of the corona, chromosphere, and photosphere at soft x-ray, XUV, and visible wavelengths, in conjunction with the observational program of the Advanced Solar Observatory.

(c) We endorse a strong program for the study of the convection zone and of the solar cycle, together with the effects of the cycle on the solar spectral irradiance and luminosity and on the heliosphere. We call attention to the need for two space missions to carry out this program:

• A Solar Internal Dynamics Mission for study of the solar convection zone and

• A Solar Coronal Explorer for study of the heliosphere.

To extend the systematic measurements of solar irradiance and luminosity initiated by the Solar Maximum Mission (SMM), we furthermore urge NASA to undertake the following:

• A well-coordinated program of synoptic observations of the solar spectral irradiance and luminosity extending over at least two solar activity cycles (i.e., the complete magnetic cycle).

(d) We support observational programs in other disciplines that will permit solar and heliospheric observations: the ISPM, OPEN, and Gamma Ray Observatory (GRO) missions; an Advanced Interplanetary Explorer; and the Solar Terrestrial Observatory.

(e) We regard the observation of chromospheric and coronal phenomena in other stars as an important part of the scientific program developed by the present Working Group. The programs of interest include Space Telescope, the Advanced X-Ray Astrophysics Facility, the Extreme Ultraviolet Explorer mission, a far-ultraviolet spectrograph in space, and future Explorer satellites.

B. Theory, Modeling, and Laboratory Astrophysics

Solar and heliospheric physics provide a unique opportunity for the theorist, since models of the basic physical phenomena occurring in an astrophysical setting can be compared with detailed high-resolution or in situ observations of fundamental plasma properties. The interplay beween detailed theoretical models and observations is therefore especially important for these disciplines and, in the university setting, for the training of astrophysics researchers.

We wish here to discuss briefly several areas of solar heliospheric theory that will be especially relevant to the scientific problems that we have identified for emphasis in the 1980's.

1. The Physics of Solar Plasmas

In the last decade, advances in plasma theory have flowed largely from studies motivated either by solar applications or by the controlled-fusion program. Among the more important developments associated with solar problems are the following:

The development of improved force-free coronal-field algorithms;

The development of magnetohydromagnetic models of coronal loops;

The study of energy release and particle acceleration by magnetic-reconnection processes, such as the static Petschek processes and the magnetic-tearing instability;

The description of reverse currents induced by electron beams and their role in heating; and

The role of electron beams in the generation of Type III radio bursts and postflare chromospheric evaporation.

During the next 10 years, the major challenge to solar plasma theorists will be the successful development of these seminal ideas in the context of the differing physical conditions and irregular geometry of the Sun, with the aim of predicting and interpreting observations. For example, the solution of the problem of the flow stability of solar magnetic structures (whether chromospheric flux tubes or coronal loops) will require a departure from the simple symmetries and static models often used. New dynamic techniques will have to be developed for this task and for the equally difficult radiative thermal instability problem. The form of the magnetic field in the solar atmosphere is the dominant influence on confinement and thermal conduction, and the relevant details of the magnetic-field architecture, such as the shear, are not at present well known in the solar atmosphere.

We anticipate that new high-resolution observations--in the coming decade with the Solar Optical Telescope and with the other major Shuttle facilities and in the early 1990's with the Star Probe--will provide important empirical guidance for the refinement of our understanding of the basic plasma processes that operate in the solar atmosphere.

2. Flares

The theory of magnetic reconnection in solar flares is still in its infancy, the considerable progress of the last five years notwithstanding. The actual time development of the Petschek process, for example, and the influence of realistic external boundaries are yet to be determined. There is not at present available any treatment of reconnection processes that includes a self-consistent calculation of the relationship of the rising current density to the turbulent resistivity. Plasma-fusion computations do not usually incorporate variable resistivity and have only recently incorporated three-dimensional geometry and such secondary instability determinants as current densities. Fully nonlinear calculations are needed to determine the spatial and temporal development of the plasma heating and acceleration in these flare models.

The study of the accelerated-particle output of a realistic flare model should receive high priority in the next decade, since it can provide answers to a number of subsidiary questions (e.g., What is the charged-particle energy spectrum? Are the conditions for the generation of a reverse current satisfied?) and predictions of observable consequences. The detailed analysis of SMM observations will result in significant progress here, but the resolving power of the Advanced Solar Observatory (ASO) will be required to provide detailed empirical guidance.

3. Coronal Heating

The mechanisms for chromospheric and coronal heating are fundamental problems that have stimulated considerable work on the basic physics of acoustic-gravity-hydromagnetic waves. The generation, propagation, and dissipation of such waves have been explored at length and an extensive body of basic physics established. More work remains to be done on various special forms of the waves and on their non-local-thermodynamic-equilibrium radiative properties, so that the forthcoming observations of atmospheric fluctuations in the chromosphere and corona can be interpreted and employed to answer crucial questions on heating. The problem is to determine from observations just which modes are involved in coronal and chromospheric heating. In addition to the dissipation of waves from the photosphere, there may be significant or even dominant heating from the dissipation of nonpotential magnetic fields. In this connection, a more complete global weak-field treatment of the conditions for (and results of) the generation of turbulence is needed. The existence of anomalous resistivity (10^5 times larger than the classical value) would have a great effect on the potentiality and rates of coronal energy-release processes and has been postulated as an explanation for rapid field reconnection and quiescent coronal heating. The availability of high-resolution observations of coronal structure at x-ray and EUV wavelengths from the Shuttle and from the Advanced Solar Observatory (ASO) is critical to theoretical studies of coronal heating.

4. Large-Scale Circulation

As noted earlier, the general hydrodynamic theory of convection and circulation in the outer envelope of the Sun is a difficult and fundamental problem. It has been demonstrated that a variety of motions in the convective core can generate the magnetic field of the Sun with the time-dependent distributions observed at the surface; however, observations are required to identify the modes that actually occur. Further progress in understanding circulation and other processes in the convection zone is thus closely linked to the Solar Interior Dynamics Program.

5. Magnetohydrodynamics of the Solar Atmosphere

Much remains to be learned about the dynamical properties of magnetic fields in conducting gases. Magnetohydrodynamics (MHD) represents a vast theoretical terrain that we have only begun to explore. Such unexplained phenomena as the general self-concentration of weak photopheric fields into separate intense individual flux tubes of 1500 gauss, and the occasional mutual gathering of those tubes into a tight cluster to form a sunspot (not to mention prominences of various kinds, spicules and surges, for example) continually remind us of the incomplete state of our knowledge of the dynamical properties of magnetic fields. A concerted MHD theoretical effort is needed to discover and work out new facets of the dynamical behavior of fields, hydromagnetic waves, neutral-point reconnection, magnetic buoyancy, the mutual hydrodynamic attraction of separate tubes, turbulent diffusion of magnetic fields, hydromagnetic dynamos, and topological dissipation. The initial flights of SOT may revolutionize the techniques used in magnetohydrodynamic models of the solar atmosphere.

6. Computer Modeling and Simulation of Coronal and Heliospheric Phenomena

We foresee a growing need for numerical studies of three-dimensional models of the magnetic fields of active regions, of three-dimensional models of the winds, fields, and cosmic-ray particle propagation in the heliosphere (particularly in connection with the International Solar Polar Mission), and of plasmas turbulence, heat transport, and rapid reconnection in models of magnetic loops, arches, and strongly sheared fields. At present, computer modeling is limited to modest Reynolds and magnetic Reynolds numbers, so it cannot explore the fully turbulent and disordered states that often occur in stellar activity. We therefore suggest that the further maturation of analytical solutions must provide the impetus for such modeling. Adequate support for modeling is needed, but at least for the present there should be no facilities created simply to handle numerical modeling whenever the request arises.

7. The Theory of Solar and Stellar Pulsations

The general theory of stellar pulsations is well developed; however, the application of this theory to the Sun is rather recent. The identification of the well-known 5-min oscillations as global p-mode oscillations of the Sun has opened up a new subdiscipline of solar physics. The general theory of pulsations in a spherically symmetric star has already been modified to allow the detection of the effects of the solar rotation and some inferences about the rotational profile of the convection zone to be drawn. As the observations of the solar oscillations become more precise, improved calculations of p-mode and g-mode oscillations will be required to test models of the internal structure of the Sun.

8. Solar and Stellar Evolution

As more precise results from neutrino spectroscopy, measurements of the solar quadrupole moment, and interpretations of the long-period solar oscillations become available, it will be necessary to consider more sophisticated models of solar evo-

lution. These should include the effects of inhomogeneous
initial-abundance models, of mixing, and of various rotational
profiles and magnetic configurations. There will be a very
active interaction between empirical results and theoretical
modeling in this subdiscipline during the 1980's.

9. Laboratory Simulation and Laboratory Astrophysics

Renewed emphasis should be placed on laboratory plasma-physics
experiments directly relevant to solar-activity problems. In
particular, experiments in toroidal geometry pertinent to
coronal-loop phenomena should be pursued; these would include
quiescent stability, end effects, turbulent heating, magnetic
reconnection, and electron acceleration. Computer simulation
by particle-pushing codes falls in this same promising category
since it provides a method of analog experimentation. However,
more effort is needed to model the complicated geometries and
nonuniformities--especially that of magnetic shear--relevant to
solar applications. The quality (and complexity) of the recent
laboratory data on solar activity furnishes a powerful motiva-
tion to theoretical developments.

   We also wish to emphasize that the availability of accurate
atomic and nuclear rate parameters, together with thermal and
nonthermal plasma models, has a profound impact on solar and
heliospheric physics, because these disciplines deal directly
with the fundamental physical processes that underlie astrophysical phenomena. We therefore urge expanded support for
laboratory astrophysics.

C. Ground-Based Observations

The resources needed in the next decade for ground-based observations can be grouped into three categories: support for
existing facilities, modifications to existing facilities, and
new facilities. We discuss these in turn.

1. Support for Existing Facilities

a. University Observatories

   University involvement in solar and solar/stellar research
is essential for continued progress in the field. A few
universities operate radio or optical solar observatories,
primarily with federal funds; these are a vital component of
the solar-physics establishment, since they serve to train new
researchers and contribute directly (and in innovative directions) to the mainstream of research. Although not strictly a

university observatory, the Mount Wilson Observatory, operated by a private foundation, furnishes important resources to the solar and solar/stellar communities, and we shall therefore include Mt. Wilson (both the solar tower and the 1.5-m telescope) in the subsequent discussion.

Funding for university observatories has failed to keep pace with the rapid inflation rate of recent years. As a result, development of new instruments has suffered and scientific staff levels have become inadequate. This trend must be reversed in the 1980's. Several observatories (e.g., McMath-Hulbert, San Fernando, Lockheed, Stanford Radio) have been closed for lack of support, while others (e.g., Clark Lake) have never obtained funds sufficient to reach a fully effective operational level. The plight of solar radio observatories is particularly acute. Only a handful remain, and several of these (e.g., Fort Davis and the University of California, San Diego facility) are in jeopardy. We recommend increased support for university observatories based on the quality of their work and the strength of their scientific proposals to funding agencies. In this connection, we wish specifically to mention the 1.5-m telescope at Mt. Wilson, which is carrying out an important program of research on stellar activity and which is in imminent danger of being closed.

b. National Astronomy Centers

Three National Astronomy Centers [the High Altitude Observatory (HAO) of the National Center for Atmospheric Research, Kitt Peak National Observatory (KPNO) and Sacramento Peak Observatory (SAO)] provide research facilities to the solar community and also conduct extensive internal research programs. It is essential to the continued progress of solar physics that these centers receive support adequate to carry out their missions. While no new major telescopes are required, there is a continuing need to improve focal-plane instrumentation and data-reduction facilities. Specifically, the continued implementation of diode-array detectors and array-processing computing facilities will be major elements of future improvements.

Long-term visitor programs have been important to all the National Astronomy Centers but these programs have now come under strong budget pressure. This trend should be reversed.

So far, sufficient time has been made available on the VLA for centimeter-wave solar observations, and this time has been effectively used to obtain two-dimensional synthesized maps of solar regions (both quiet and active) with a spatial resolution of 1 arcsec or less. However, we believe that the effectiveness of the VLA for solar observations would be greatly enhanced by the establishment of an internal solar program at the National Radio Astronomy Observatory. This could be accomplished through

the addition of one or two solar radio astronomers to the staff.
Adequate time should also be provided on the VLA for observations of solar-type stars.

c. The Homestake Mine Neutrino Experiment

This experiment, based on chlorine, should be operated
concurrently with a second experiment, now under development,
that will utilize substantial quantitites of gallium. This
approach will permit two populations of neutrinos (from different branches of the proton-proton chain) to be sampled
concurrently.

d. Image-Reconstruction Technique Development

Postobservational reconstruction of optical images degraded
by atmospheric seeing has great potential for solar physics.
Several institutions are exploring alternative methods. If any
of these becomes practical, resolutions of 110 km and 40 km
could be achieved with existing telescopes at SPO and KPNO,
respectively. Thus, ground-based telescopes could compete in
performance with space telescopes for certain limited objectives, at considerable cost savings. The Working Group urges
continued support for these developments.

2. Modifications to Existing Facilities

a. The Very Large Array

Coordinated observations between the ASO and the VLA for
the study of the coronal magnetic field, the acceleration and
propagation of electrons during the impulsive phase of flares,
and coronal transients will be an important part of solar-
physics research during the 1980's. The VLA should therefore
be better adapted to the needs of solar observations. In
particular, the VLA capability to measure circular polarization
should be improved to an accuracy of better than 1 percent, and
the technical problems related to dynamic range and calibration
of total solar-flux measurements should be solved. It is
encouraging that some of these modifications are currently
under way and that others are planned.

b. Solar Radio Observatories

Although the VLA provides important and unique capabilities
for solar radio observations, many solar problems require a
dedicated facility capable of undertaking long-term studies of
coronal magnetic fields and of transient and cyclic phenomena
in the corona and chromosphere in collaboration with ground-

based and space observations. Among the phenomena that can be studied are particle acceleration in flares (through observations of Type III bursts), wideband spectroscopic observations (such as are possible at Fort Davis and Owens Valley), and polarization studies of phenomena in the outer corona (such as are possible at Clark Lake) to determine the configuration of the coronal field. It is important that improvements be made in these facilities to make them more effective and more accessible to visitors, which will require an augmentation of funding for instrumentation. Among the projects discussed by the Working Group, we wish to mention in particular the use of array processors for interferometric facilities, improved computing facilities, and additional antennas for improved polarization measurements.

c. South Pole Solar Observations

Solar observations made from the South Pole station in the Antarctic summer of 1978 showed great promise for this site as a cost-effective means of studying solar phenomena that require long periods of uninterrupted observation. The use of the South Pole for long-term observations of the solar oscillations is an integral and key component of the program for the study of the convection zone and of the activity cycle outlined earlier. This international observing station should be improved by providing a modest telescope capable of supporting a wide variety of instrumentation. A program of guest investigations should be established to make maximum use of the unique opportunities offered by the site.

d. Stellar Magnetic Fields

In view of the central role played by magnetic fields in active phenomena in the solar atmosphere, a program of greatly improved observations of stellar magnetic fields is important in order to improve our understanding of active stellar phenomena in general. Recent instrumental advances suggest that major improvements in both magnetic and velocity-field detection can now be applied to stellar observations relevant to improving our understanding of extended stellar atmospheres and of activity cycles in stars.

e. National Astronomy Centers

Improvements at the National Astronomy Centers in telescope performance and efficiency (for example, through the use of array detectors), data-reduction equipment, and computing facilities should be adequately supported during the 1980's. Examples are the planned improvements in guiding for the McMath telescope at KPNO and the Big Dome instrumentation at SAO.

Such improvements are fundamental to the achievement of the improved performance necessary to take full advantage of the sophisticated focal-plane instrumentation discussed below.

3. New Equipment and Programs

In order of priority, these are as follows:

   a. Neutrino Experiments

   The gallium neutrino experiment currently under development promises fundamental new information on the rates of thermo-nuclear reactions in the Sun. We cannot think of an experiment that is more likely to influence the course of stellar physics in the 1980's. We also hope that the lithium and indium exper-iments discussed earlier can be initiated in the 1980's.

   b. Velocity-Detector Development

   A new generation of sensitive, stable detectors for the measurement of small velocity shifts is being developed for studies of global circulation, long-period oscillations, and convection. Among these are the Fourier tachometer, resonance absorption cells, a heterodyne spectrograph, stabilized spec-trographs, and narrow-band filters and interferometers. The funds required are modest in comparison with the value of the anticipated scientific benefits. These developments are a key component of a long-range program for the study of the convec-tion zone and the solar activity cycle and will provide the prototypes for instrumentation on a Solar Interior Dynamics Mission.

   c. Stellar Telescope

   A stellar telescope--dedicated to synoptic studies of stellar activity analogous to solar activity and preferably located at a National Astronomy Center--is needed immediate-ly. A moderate aperture (about 2.5 m) is sufficient, but a modern high-resolution spectrograph, digital two-channel photometer, and fast accurate pointing under computer control are essential. A proposal to build such facility at SPO with significant Air Force funding is at present under study by the Association of Universities for Research in Astronomy and NSF. We hope that this project will go forward.

   d. Real-Time Image Restoration

   The development of active-optics devices for real-time image restoration is proceeding rapidly. We anticipate that these devices will shortly become available for solar applications that require diffraction-limited performance at solar telescopes. We urge that KPNO and SPO be granted additional funds to purchase these expensive devices as they become available, for the benefit of their visiting scientists.

e. Vector Magnetograph

Nearly all of our observational knowledge of solar magnetic fields has been derived from measurements of the line-of-sight component of the field in the photosphere. With a vector magnetograph, this scalar view can be extended to a more complete vector picture that will also allow electrical current systems to be determined. Recent instrumental and interpretational advances, together with the possibility of comparing high-resolution SOT measurements from space with synoptic ground-based measurements of lower resolution, offer exciting prospects for constructing a vector magnetograph capable of greatly furthering our understanding of solar magnetic fields. Because the construction cost will be substantial, this instrument should be installed at a National Astronomy Center in order to serve the largest number of visiting scientists.

f. Measurements of the Solar Diameter

Precise measurements of the solar diameter can provide information on a number of fundamental properties of the Sun, including the solar oblateness (hence internal rotation) and long-period oscillations. We suggest that more sensitive instrumental approaches should be sought and supported to furnish a better understanding of the solar interior.

g. Solar Electrograph

There are several possibilities for determinating transient electric fields in the solar atmosphere through observations of the Stark effect in hydrogen and helium. Among these options is a new instrument designed to measure the asymmetric linear polarization produced across a Balmer line profile. Advances in optics, detectors, and electronics offer the possibility of constructing an electrograph that would allow measurement of a physical parameter of critical importance in flare studies and other aspects of activity.

4. Solar Monitoring

Several of the major problems of solar physics require regular observations over long periods of time for effective progress. Facilities for conducting some of these observations from space

may not be available until the 1990's. Even then it may prove
to be more cost-effective to conduct synoptic or monitoring-
type observations from the ground. The need for continuous
monitoring of certain solar phenomena is illustrated by the
"discovery" of the nearly rigid rotation of coronal holes
through analysis of space observations acquired during
1972-1973. Continued monitoring of coronal holes from the
ground from 1974 onward, however, showed that this apparent
rigid rotation was actually an anomaly associated with that
phase of the solar cycle. A second example is the discovery of
torsional waves in the latitude dependence of the differential
rotation from data obtained over a significant fraction of the
activity cycle at Mt. Wilson. Numerous other examples of the
inadequacy of too-brief or snapshot investigations of solar
phenomena can be cited. A careful program of monitoring of
selected solar phenomena must be continued to prevent any bias
in our view of the Sun introduced by a restriction in the range
of observations. Both university observatories and the National
Astronomy Centers must play an important role in this activity.

D. Suborbital Programs

Rocket and balloon flights can be especially valuable for solar
physics: the Sun is sufficiently bright to allow extremely
productive observations in a short time, and many solar observations require the development of highly specialized observational instrumentation for time-limited but important observational programs.

These low-cost vehicles thus serve a dual purpose for solar
physics, allowing significant and timely scientific programs to
be carried out at modest cost and allowing the development of
sophisticated new instruments in a cost-effective way. We
believe that rockets and balloons will continue to furnish
important observational opportunities for solar physics into
the indefinite future and will become even more effective with
the development of the Experiments of Opportunity Program (EOP)
to place rocket payloads in temporary orbits from the Shuttle
and the development of long-duration balloon flights. We regard
the continuation of a strong rocket and balloon program, including the development of the capability for the extended observations just mentioned, as part of the foundation of solar-physics
research from space.

E. The Space Shuttle

The effective use of the Space Shuttle is the core of the
scientific program developed by the Working Group because it
will allow effective scientific programs to be carried out

expeditiously on the Shuttle itself, while simultaneously permitting the evolutionary development of instrumentation that can be adapted for long-term missions with a minimum of modifications. Observations from the Shuttle, even with the initial limitation of 7 days in orbit and the further constraints imposed by a multipurpose flight, will be extraordinarily productive for solar physics. Therefore, instrumentation for ultimate placement on three of the most important solar space missions--the Advanced Solar Observatory (ASO), the Solar Coronal Explorer (SCE), and the Solar Interior Dynamics Mission SIDM)--can be initially developed and used to carry out a significant scientific program within the framework of early Shuttle sortie missions, permitting an evolutionary program of high productivity and cost-effectiveness.

We here divide our discussion of Shuttle instrumentation into three sections: the development of facility-class instruments to form eventually a Shuttle Solar Observatory, the prototype for an ASO; a program of observations of the solar oscillations from a polar-orbiting Shuttle mission as a part of the Solar Cycle and Dynamics Program; and the development of PI instruments for specialized observational programs and as prototypes for those on ASO and SCE.

1. A Solar Shuttle Observatory

The Working Group supports the evolutionary development of a Solar Shuttle Observatory (SSO), composed of several advanced facility-class instruments, to be used in a coordinated group on the Space Shuttle. (The term, "Shuttle Solar Observatory," is used here to denote the ensemble of solar facility-class instruments that will eventually form the ASO; it does not imply a program distinct from the programs to develop the individual component facilities, such as SOT and the Solar X-Ray and EUV Facilities) The development of these instruments should proceed as part of the ongoing Spacelab program, with the SOT being the first instrument to be developed. The solar-physics community has identified a Solar Soft X-Ray Telescope Facility (SSXTF) and a Pinhole/Occulter Facility (for hard x-ray imaging and for the study of the corona at high angular resolution) as additional major components of the SSO. Other major SSO instruments that have been identified by the Solar Shuttle Facility Definition Teams include facilities for EUV and gamma-ray observations, a wide-field optical facility for the study of solar oscillations, and radio spectrographs and polarimeters to observe plasma processes in the corona and the solar wind. It is possible that some facilities can be developed initially as PI instruments; the SSO might also include specialized PI instruments to carry out a variety of coordinated observing programs. The European Space Agency

(ESA) is currently studying a Grazing Incidence Solar Telescope (GRIST) for XUV observations. Should ESA decide to develop this instrument, it also could become part of the SSO.

The SSO will form the nucleus of an Advanced Solar Observatory (ASO), which can be established by refurbishing the SSO instruments and placing them on a space platform. The instruments that appear to have the greatest scientific promise in such a role are discussed below.

a. The Solar Optical Telescope

The SOT is a Gregorian telescope with an aperture of 125 cm, which will be designed to provide a resolution of 0.1 arcsec at 5500 Å (72 km on the Sun) and 0.02 arcsec (with image restoration) at 1100 Å SOT will operate from 1100 Å into the infrared and can thus observe solar phenomena from the photosphere up through the chromosphere and transition region to the base of the corona. The telescope will accommodate a number of PI-class focal-plane instruments to be chosen by peer evaluation but that are likely to include narrow-band filter magnetographs, very-high-resolution visible and ultraviolet spectrographs, and a Stokes polarimeter, with possibilities for simultaneous operation of two or more of these instruments. The basic structure (cannister) of the telescope can accommodate several additional PI- or facility-class instruments that provide their own collecting optics in a single cluster and that can therefore form the nucleus of the Shuttle Observatory.

The basic scientific requirement for SOT is the need (discussed elsewhere in this chapter) to achieve the very high spatial resolution required to determine the density, temperature, magnetic field, and nonthermal velocity field in a large number of solar features on the small scales characteristic of the basic physical processes involved. These processes include changes in the strengths of highly localized magnetic fields, small-scale turbulent velocity fields, hydromagnetic waves and pulses, and systematic mass flows--all important for an understanding of the overall structure, dynamics, and energetics of the solar atmosphere. To study such processes, it is necessary to resolve regions over which significant gradients exist in the local magnetic and velocity fields, as well as in local densities and temperatures. Observations must be made on the scale at which the material "clumps." Recent observations suggest that the plasma in the solar chromosphere and lower transition region often varies significantly on a scale of a few tenths of an arcsecond. In addition, 0.1 arcsec is approximately the mean free path of a typical photon in many of these structures, so a natural limit to remote sensing of the solar atmosphere will be achieved by this telescope.

b. Solar Soft X-ray Telescope Facility (SSXTF)

The SSXTF is a 0.8-m nested Wolter Type I telescope configured to provide efficient imaging of x rays from about 1.7 Å (the wavelength of the important lines of Fe XXV and Fe XXVI) to 300 Å. An effective collecting area of some 1200 cm^2 over most of the wavelength range, together with an angular resolution of 0.3 arcsec are design goals. Detailed studies of candidate focal-plane instruments for SSXTF are currently under way and indicate that the use of grazing-incidence focal-plane optics and stigmatic spectrometer configurations will provide performance charateristics that will allow this instrument to achieve a tenfold gain in high spatial and temporal resolution spectrohelioscopy and in the study of line profiles, by comparison with any previous solar soft x-ray instrument. As pointed out earlier, the thermal gradients in coronal-loop structures, possible thermal enhance-ments in the corona associated with coronal current sheets, the fine structure of coronal bright points, and the impulsive energy release and transformation that accompany a flare all occur on a scale smaller than the 2-3 arcsec resolution achieved with broadband filtergrams in previous long-duration observations (one or two rocket observations have recorded phenomena at the 1 arcsec level with limited sensitivity). SSXTF will be able to obtain high spectral resolution and to study line profiles at a resolving power that should allow the detailed physical processes operating in the corona to be stud-ied. SSXTF would initially be deployed on Spacelab in the SOT cannister described above and would later form part of an ASO as discussed below. A detailed engineering definition study for the SSXTF has already been carried out. No major technical problems in its development are foreseen, although the mirror fabrication will constitute a significant challenge because of the requirement for high angular resolution.

c. Grazing Incidence Solar Telescope (GRIST)

The GRIST is an XUV facility currently being considered for development by ESA. It will be capable of obtaining EUV and XUV spectra between 100 and 1500 Å through use of grazing-incidence and normal-incidence spectrographs fed by a Wolter Type II grazing-incidence mirror. The spatial resolution will be 1 arcsec and the spectral resolution in the range 1×10^4 to 3×10^4. It is a powerful instrument that could nicely complement SOT and SSXTF if it can be configured to be flown on the Shuttle or on a platform simultaneously with SSO and/or with an ASO. Definition work already performed indicates that development of this facility is feasible; at the time of this report, however, ESA has not decided whether this facility will be funded for development.

d. EUV Telescope Facility (EUVTF)

The EUV will consist of a normal-incidence telescope
feeding stigmatic EUV spectroheliographs operating between 400
and 1500 Å. The design objective for spatial resolution is
0.1 arcsec, and that for spectral resolution lies between 10^3
and 10^5. This instrument will complement the SOT by extending
ultra-high-resolution observations to radiation characteristic
of temperatures as high as 2×10^6 K. It is intended to
achieve a tenfold improvement in angular resolution and spectral
resolution by comparison with any currently approved instrument
that is capable of observing material on the solar surface
hotter than some 2×10^5 K.
Study of the fine structure of material at temperatures
between those of the chromosphere (about 10^4 K) and the corona (greater than 10^6 K) will certainly require resolution on
the order of 0.1 arcsec. Such fine structure includes the
chromospheric network, which is known to extend to material at
5×10^5 K, coronal bright points, the thermal gradients in
coronal loops, and the structures responsible for the prompt
EUV bursts accompanying the impulsive phase of a flare.
A preliminary design study for a solar EUVTF has already
been completed by the Solar Shuttle Facility Definition Team.
The EUVTF could be accommodated within the SOT cannister
simultaneously with the SSXTF.

e. The Pinhole/Occulter

The Pinhole/Occulter is a unique observing facility that
will make use of a remote occulting mask to obtain high-resolution images. In principle, there is no limit to the angular
resolution that can be obtained in x radiation, since the
resolution is a function only of the separation of the detector
and mask. Coded-aperture systems such as the Pinhole are most
effective at hard x-ray energies (above 5 keV). The remote
mask will also simultaneously be used as an external occulting
disk for large-aperture coronographic telescopes to permit the
first high-resolution studies of coronal structure that is close
to the solar limb and in the region of solar-wind acceleration.
Observations of EUV emission lines and soft x radiation should
be possible in the far corona with the large-aperture optics
made possible by the remote occulting mask.
A preliminary study of a Pinhole/Occulter Facility is
currently in progress at NASA. An initial Pinhole/Occulter
system for Spacelab might have a baseline separation of 50 m
with a mask mounted on an arm extended from the Shuttle orbiter;
it would achieve resolving power of a few tenths of an arcsecond
at 10 keV. Higher resolution can be achieved by placing the
mask and detector on separate vehicles. For example, the
aperture mask might be carried on the orbiter (together with

the SSO) and the detector carried on a separate subsatellite placed in temporary orbit by the Shuttle.

The Pinhole/Occulter concept is at the core of the flare-oriented aspects of the scientific program of the SSO/ASO, because details of the reconnection and acceleration processes can only be studied directly with high-resolution (0.1 arcsec or less) hard x-ray observations also yielding good time resolution.

f. Solar Gamma Ray Telescope (SGRT)

The acceleration of nuclear particles to high energies and their propagation from the acceleration region to the inter-planetary medium is one of the least well understood flare phenomena. It has long been thought that this acceleration occurs in two stages, with initial acceleration to perhaps a few hundred keV to 1 MeV occurring simultaneously with the impulsive acceleration of electrons and acceleration of part of the population of high-energy nuclei to ultra-high energies (up to GeV per nucleus or more) occurring later. Data from SMM have apparently confirmed that protons are accelerated in the initial phase of the flare. In order to study these processes, simultaneous hard x-ray and gamma-ray observations are requir-ed, since arcsecond gamma-ray observations are not feasible even with the Pinhole/Occulter.

A Solar Gamma-Ray Telescope for flare observations must cover the energy range from 100 keV to 10 MeV, in order to observe both the bremmsstrahlung associated with the impulsive acceleration and the nuclear gamma-ray lines associated with both acceleration processes. The SGRT must have high resolu-tion at MeV energies to resolve lines and detect Doppler shifts and so must utilize cooled-germanium-detector technology. An instrument having the general characteristics of the HEAO-3 gamma-ray spectrometer would be ideal. Although we support the development of the SGRT for initial Shuttle use, its main solar observational program can only be carried out on a long-dura-tion ASO.

g. PI Instruments

The development of PI instruments is an important component of the SSO Program for two reasons. First, the sequential pro-cedure that must be used to develop the various SSO component facilities makes it highly desirable to use PI soft x-ray, XUV, and coronal intruments to accompany the early flights of the SOT until the SSXRT, EUVTF, and Pinhole/Occulter Facilities become available. Second, several instruments that will form an integral part of SSO and ASO are small enough to allow their development as PI instruments rather than as facilities. Certainly, in these cases, the PI approach is the preferable one. Among the instruments that fall into this category are white-light and ultraviolet coronagraphs (for use with the Pinhole/Occulter), radio spectrographs and polarimeters, and wide-field optical instruments for studies of the solar oscillations. We later discuss the individual PI instruments reviewed by the Working Group in more detail.

2. The Study of Solar Interior Dynamics from the Shuttle

A sequential program for the study of the internal dynamics of the convection zone has been discussed earlier. This program envisages the development of instrumentation for the observation of the solar velocity and magnetic fields and for precision differential radiometry of photospheric and chromospheric structures; observations would begin on the ground at the South Pole of the Earth and be followed by Space Shuttle observations employing PI-class instruments from a fully sunlit polar retrograde orbit. The timing of this observational program is a critical step in the development of a Solar Interior Dynamics Misson; we urge NASA to share with NSF the development costs of Fourier tachometers and precise differential radiometers and to initiate the development of instruments intended for Shuttle use as soon as possible.

3. The Shuttle-Attached Solar Terrestrial Observatory (STO)

The STO is a Shuttle facility intended for the study of the Earth's atmosphere and ionosphere and of the solar radiative and particulate fluxes that influence it. STO is currently under study at NASA and would be operated initially as a Shuttle-attached facility. STO will observe the solar spectral irradiance, the solar total radiative flux, and the solar wind through observations of the large-scale corona, as well as study the terrestrial atmosphere and ionospheres. The observing program of STO is thus highly complementary to the Solar Interior Dynamics (SID) program; STO will emphasize the coupling of the solar radiative flux and the solar wind (via the heliosphere) to the Earth's atmosphere, whereas the SID program will emphasize the coupling between the solar radiative flux and the solar wind and the structure and dynamical behavior of the Sun's convection zone and the corona.

4. PI Instruments

Several PI instruments that will be an integral part of the observational program that we have developed have already been identified through the first Shuttle instrument selection and various definition studies. We briefly review these instruments below:

(a) The High Resolution Telescope and Spectrograph (HRTS), which operates between 1100 and 1800 Å with 1-arcsec resolution and high spectral resolution, is under development for Spacelab II. It is a powerful diagnostic instrument for the study of the chromosphere and transition region.

(b) White-light and Lyman-alpha coronagraphs. This pair of experiments, successfully flown on a rocket, has provided for the first time a technique to measure coronal temperatures and mass-flow velocities between 1.5 and 4 solar radii. The technique uses measurements of the intensity and profile of the resonantly scattered hydrogen Lyman-alpha line and the intensity and polarization of the electron-scattered white-light corona. It is a powerful experiment package that could greatly expand our empirical knowledge of the physical state of the coronal plasma beyond 1.5 solar radii in the region where the solar wind is accelerated.

(c) Soft x-ray telescopes. Several different imaging soft x-ray telescopes and spectrometers have been or could be developed to provide high-spatial-resolution imaging (1 arcsec) with broadband filters, or moderate-spatial-resolution imaging (10 arcsec) with a spectral resolution of about 10^3. Such instruments could provide a nice complement to SOT prior to the development of the SSXTF as well as perform important studies of solar activity, coronal holes, flares, and coronal brights points, for example.

(d) EUV/XUV telescopes. A variety of different types of EUV and/or XUV telescopes and associated spectrographs have been or could be developed to provide EUV and XUV spectra and spectroheliograms with good spatial resolution (1-3 arcsec) and a spectral resolution in the range 10^3-10^4 with moderate temporal resolution. Although smaller than EUV/XUV facilities, such PI instruments can still carry out highly valuable scientific programs and provide cost-effective methods for attacking selected scientific problems operated either alone, in conjunction with other PI instruments, or with SOT and/or the SSXTF.

(e) Low-frequency radio spectroscopy and polarimetry. Dynamical processes occurring in the corona and in the solar wind can be studied through low-frequency (decametric to kilometric) observations from space. Both spectroscopic and polarimetric observations are needed. This observational program can elucidate the interaction of nonthermal plasma reams in the corona, collisionless shockwaves, and shock-induced particle acceleration; it will furthermore be highly complementary to in situ observations from ISPM, the Interplanetary Laboratory, and Star Probe.

F. Explorer Missions

The scientific program defined by the Working Group includes
two scientific objectives that can be achieved by Explorer-
class spacecraft. A SCE could carry out high-resolution,
in-ecliptic observations of the corona at x-ray, XUV, and
visible wavelengths; a Sun-synchronous SIDM could study the
structure of the convection zone through observation of the
global oscillations of the Sun for extended periods.

1. Solar Coronal Explorer (SCE)

An essential element in understanding both the short- and
long-term effects of solar activity on the interplanetary
medium is the documentation of the three-dimensional structure,
dynamics, and evolution of the corona and interplanetary medium
and the origin of the solar wind. Both in situ observations of
the interplanetary plasma, such as will be provided by the
International Solar Polar Mission (ISPM) and the Interplanetary
Physics Laboratory (IPL), and synoptic observations of the
global configuration of the coronal magnetic field, coronal
density structure, coronal temperature distribution, and solar
wind flow velocity distribution will be required. The necessary diagnostic studies of the temperature density and flow
velocities of the transition region and very low corona are
most accurately carried out by a high-resolution EUV spectrograph, while the configuration of the magnetic fields, temperatures, and densities in the low corona are most reliably
established by soft x-ray images. In the intermediate and
outer corona, the density, temperature, and velocity may be
established by observations of the continuum corona (which
yields the distribution of density) and by observations of the
intensities and profiles of resonance lines scattered in the
corona (which reflect the temperature and expansion velocity).

We believe that a SCE should be implemented as soon as
possible. NASA has established a Science Working Group to
investigate in detail how the objectives of SCE might be met,
what instruments should be carried aboard such a mission, and
how they might be accomplished within the constraints of
Explorer-class missions.

Although each mission has a coherent individual scientific
program, SCE, ISPM, and IPL will, if flown simultaneously,
greatly enhance the scientific yield of the program that we
have outlined. Observations from SCE and ISPM can be combined
to yield stereoscopic observations of the corona, from which
its three-dimensional density structure can be inferred, and
observations from ISPM and IPL can be combined to yield
simultaneous in situ observations from two locations in the
interplanetary medium widely separated in latitude. These

three missions should operate together for at least a year
bracketing the polar passage of ISPM.

2. Solar Interior Dynamics Mission (SIDM)

Our program requires the use of observations of photospheric
oscillations, large-scale circulation, and global radiative-
flux variations to establish fundamental parameters and
boundary conditions for theories of the solar cycle. SIDM will
provide both the first opportunity for high-frequency-resolution
observations of solar oscillations and the first "look" deep
into the convection zone of the Sun.

    Several instruments carried aboard a full-sunlight, orbit-
ing SIDM are required to meet these objectives. As described
earlier, simultaneous, low-spatial-resolution observations of
photospheric velocity and magnetic fields will provide the
basis for study of solar oscillations and large-scale circulation patterns, together with their relation to magnetic fields.
Likewise, precision differential radiometry will reveal the
temperature signature of large-scale circulation cells.
Precision total-radiative-flux measurements are required to
investigate the possible feedback between solar magnetic
activity and solar flux and the possible consequence of
long-term flux variations on terrestrial climate. Finally,
spectral-irradiance measures will provide insight into changes
in the structure of the solar atmosphere and the consequences
of spectral-irradiance variations in the terrestrial atmosphere.

G. Major Space Missions

We have identified two major missions that will require either
a specialized spacecraft or a space platform. The first, an
ASO, may be developed by placing the SSO on a long-duration
Space Platform. The second, the Star Probe, can place an
instrumented package into orbit that will allow a close
encounter (within about 3 solar radii) with the Sun. We also
discuss a closely related solar-terrestrial mission, the STO,
which will have a major impact on solar and heliospheric
physics.

1. Advanced Solar Observatory

The SSO can effectively address those scientific problems that
can be attacked with measurements obtained over periods of
hours to days with selected instruments. However, other
classes of problems require observations made over longer
periods of time--weeks to months--because of (a) the

infrequency of the phenomena (e.g., large flares, eruptive prominences); (b) the evolutionary nature of the phenomena (e.g., active regions, coronal holes, large-scale coronal structures, etc.); and (c) the need for synoptic data to address the problem properly (e.g., a detailed probing of the three-dimensional structure of the corona beyond 1.5 solar radii through measurements of optically thin radiation with coronagraphs).

For example, flares are a complex phenomenon reflecting numerous interrelated processes that produce radiation across the electromagnetic spectrum from radio to gamma-ray wavelengths. All of these radiations must be used to probe the physical processes responsible for the flare phenomenon, which involve plasmas having temperatures ranging from 4×10^3 to 10^8 K and densities from 10^4 to 10^{21} particles/cm^3. The necessary observational programs can be performed most effectively on a free-flying platform serviced by the Shuttle because of its versatility, capability for instrument modification, and long operational lifetime. A platform also provides superior thermal control, a contamination-free environment, and a more stable base for high-resolution observations than the Shuttle can provide. A Shuttle-serviced platform can also provide the flexibility of periodic retrieval, refurbishment, or replacement of major components of ASO.

a. Scientific Objectives

An Advanced Solar Observatory will provide an opportunity for a broad frontal attack on nearly all the major problems identified by the Working Group. The first ASO mission could make use of a battery of instruments to investigate simultaneously the thermal energy structure and dynamics of the photosphere, chromosphere, and corona. Of particular interest is the physical state of the solar atmosphere from the photosphere through the corona to beyond 5 solar radii from the Sun's center, which is critical to an understanding of the physical nature of the solar atmosphere. A later mission of the ASO, centered on the 1991 solar maximum, could observe flares and other phenomena that require very high spatial resolution (0.1 arcsec) at optical and x-ray wavelengths and high-spectral-resolution gamma-ray measurements.

Recent observations have begun to indicate that small-scale structures, such as magnetic-flux bundles, pinched currents, and filamentary emission regions, form the basic components of active phenomena in the Sun. Such observations are consistent with the predictions of certain theories--for example, those that postulate microinstabilities leading to "anomalous resistivity" in current interruption or filamentation models of flares. We will, therefore, require observations with higher spatial and temporal resolution (especially at x-ray and gamma-ray wavelengths) to clarify mechanisms and seek ultimate causes of active phenomena. ASO is the only facility capable of such an observational program.

ASO will also provide the means to continue the studies of the dynamics of the convection zone and the structure and dynamics of the corona to be begun with SIDM and SCE. These studies must cover the full 22-year solar magnetic cycle, if the relationship between the dynamical behavior of the convection zone and the magnetic and activity cycles is to be understood. Collaborative observations between ASO, the Star Probe, and the VLA will furthermore be important to each of the scientific problems that we have discussed.

b. Development Plan

The evolutionary development of Shuttle-attached instrumentation to form a SSO offers the chance to procure and test instrumentation suitable for ASO in a cost-effective and scientifically productive manner. The broad range of solar observations possible with most of the necessary instrumentation will ensure its effective, coordinated use on brief, early Spacelab missions. ASO should take its final form by 1988 at the latest, implying a formal "new start" by 1983, with refurbishment in 1990 to take advantage of the solar maximum for flare studies.

c. ASO Configurations

The instrument complement envisaged for the first ASO flight would ideally include the SOT, the Soft X-Ray, and EUV facilities, and a scaled-down version of the Pinhole/Occulter. We believe it imperative, however, to initiate ASO as soon as possible; therefore, if the development of Shuttle facility-class instruments must proceed slowly, the initial flight of ASO should use PI-class instruments rather than be delayed. For example, we can envisage a first ASO flight containing SOT, SSXTF, and EUV, XUV, and coronagraph instruments in the PI class. Depending on how the development of a space platform proceeds, initial ASO flights might be as brief as 6 months; however, we believe that, following full operation, at least 30-month flights with 6-month intervals for refurbishment should be the goal. The ASO mission that occurs during the next solar maximum should include the Pinhole/Occulter, the Gamma Ray Telescope, and radio instrumentation in addition to the nucleus of instruments described above.

2. The Star Probe

The Star Probe is a proposed mission that would penetrate to within 4 solar radii of Sun-center, passing over the poles of the Sun during the declining phase of a solar cycle. This spacecraft would make in situ observations of the interplanetary medium, remote observations of the solar atmosphere, and an inference of the solar mass and angular-momentum distribution from precise (drag-free) orbit determinations. The primary scientific objectives of the Star Probe mission include an improved understanding of (a) the energy balance of the solar corona and acceleration of the solar wind; (b) the distribution of mass and angular momentum in the solar interior; (c) the three-dimensional fine structure of solar magnetic elements and coronal loops; and (d) the physical processes governing the evolution of the solar magnetic field and the acceleration, storage, and release of solar energetic particles. This mission would be truly exploratory in its in situ investigation of the innermost heliosphere and its ultra-high-resolution (7-10-km) images of the structure of the solar atmosphere. It would be expected to reveal phenomena not yet considered in studies of the Sun and its atmosphere. A polar orbit is preferred for the Star Probe trajectory to permit sampling of the latitude dependence of conditions in the solar-wind acceleration region. The timing of the Star Probe mission is critical, since solar encounters should occur at solar minimum to maximize the probability of traversing polar coronal holes.

The Star Probe, while a very demanding mission, will allow unique observations that directly address major scientific questions of stellar physics, together with a tenfold improvement in resolving power over most of the electromagnetic spectrum compared with the resolution that can be achieved in Earth orbit, even with ASO. The Star Probe would also allow unique stereoscopic observations of coronal structures in collaboration with the ASO, both during its cruise phase and during encounter.

3. The Solar Terrestrial Observatory (STO)

The Shuttle-attached STO described earlier should evolve into a space platform-supported mission for investigation of the mechanisms that couple solar activity to the terrestrial system. Such a facility would incorporate solar, magneto-spheric, and atmospheric instruments to observe solar structure and variability, together with the influence of this variability on the terrestrial magnetosphere and atmosphere, in a simultaneous, coordiated manner over extended periods of time. It is planned that the platform operation of STO would be accompanied by the operation of an interplanetary space mission, such as the Interplanetary Laboratory of OPEN or the Advanced Interplanetary Explorer (AIE) mission, both of which are discussed below. The major solar objectives of STO are to establish the intrinsic variability of the radiative and

particle fluxes from the Sun, the physical characteristics of
the solar phenomena responsible for these variations, and the
relationships between these solar phenomena and the resulting
variations in photon flux, particle flux, and magnetic field in
the vicinity of the Earth.

These objectives of the STO dictate an instrument ensemble
that is fundamentally different from that envisaged for the SSO
and its platform-supported successor, the ASO. The STO instruments will all be low-resolution instruments with full-Sun
coverage, rather than the high-resolution instruments with
limited field of view appropriate to the SSO/ASO scientific
objectives. A strawman STO payload has been defined for two
configurations. The solar part of the basic STO configuration
includes solar-constant and -irradiance monitors (at UV and EUV
wavelengths), a full-Sun, soft x-ray telescope, and both white-
light and resonance-line coronagraphs. For the advanced configuration, a full-Sun XUV Doppler spectroheliograph and a
full-Sun hard x-ray spectrometer are to be added to the basic-
configuration instruments. This ensemble of instruments will
determine the properties of solar events that lead to terrestrial consequences and are aimed at providing a solar/ter-
restrial predictive capability through an understanding of the
mechanisms that play a role in the Sun and in the solar corona.

4.   In Situ Observations of Heliospheric Phenomena

We wish to emphasize, in the strongest possible way, that in
situ observations of the properties of the interplanetary
medium constitute an _integral_ part of the scientific program of
solar/heliospheric physics. We therefore discuss here the in
situ missions most closely connected with the scientific program outlined earlier.

    a.   The International Solar Polar Mission (ISPM)

    ISPM represents the first exploration of the three-dimen-
sional structure of the heliosphere outside the ecliptic plane.
The originally approved ISPM mission profile provided for two
spacecraft, aided by a Jupiter gravity assist, to enter symmetrical polar orbits around the Sun with a perihelion distance of approximately 1 AU, with each spacecraft to measure
particle fluxes and magnetic fields, and observations of coronal structure in x rays, XUV, and white light, of hard x-ray
bursts, and of dust to be made by at least one spacecraft. As
noted near the beginning of this chapter, the Working Group
believes that the implementation of ISPM as a dual-spacecraft
mission with imaging capability is essential to an orderly and
effective program of solar and heliospheric studies; simultaneous observations in the northern and southern hemispheres of

the heliosphere is a unique and basic feature of the ISPM
concept and is critical to achieving an understanding of the
global structure of the heliosphere. The imaging experiments
on ISPM will also provide new insights into the structure of
coronal holes, the structure of coronal streamers, and the properties of coronal transients.

b. The Interplanetary Laboratory (IPL) of OPEN

IPL will provide a platform for studies of the interplanetary medium, solar-wind plasma energetic particles, and the
heliospheric magnetic field in the ecliptic plane; it will also
provide the collaborative observations required to understand
the dynamical behavior of the Earth's magnetosphere, which will
be observed by the three additional OPEN spacecraft within the
magnetopause. In collaboration with ISPM and SCE, IPL will
provide a unique opportunity to study the dynamical behavior of
the heliosphere.

c. An Advanced Interplanetary Explorer (AIE)

The objective of the AIE mission is to study the flux, isotopic composition, and propagation of solar and Galactic cosmic
rays from hydrogen up to iron, both inside and outside the
heliosphere. An understanding of the acceleration and propagation of solar cosmic rays is a fundamental objective of solar
and heliospheric physics. AIE could operate simultaneously
with Star Probe and with ASO, so that a coordinated attack on
the problems of the acceleration and transport of cosmic rays
in the corona, and the influence of the three-dimensional
structure of the heliosphere on their subsequent propagation,
could be undertaken.

5. The Study of Active Stellar Phenomena

The study of stellar activity, like the study of solar activity,
will require a coordinated program that permits the simultaneous
observation of stellar emission from optical to x-ray wavelengths for investigations of the interrelationship of stellar
magnetic fields, "star spots," and chromospheric and coronal
phenomena. Space Telescope, an Advanced X-Ray Astrophysics
Facility, and a far-ultraviolet spectrograph in space will all
have substantial capabilities for the study of active stellar
phenomena and should play significant roles in this area.
Since the Earth appears to be in a region of relatively low
interstellar-medium density, a substantial number of stellar
sources can also be studied at EUV wavelengths; an orbiting
soft x-ray/EUV observatory would be an enormously powerful
instrument for such work.

VII. THE SOLAR-PHYSICS COMMUNITY

Although solar physics has maintained its vitality during the
past decade, as the preceding sections of this chapter demonstrate, the community of solar physicists faces a number of
major problems, not all of which can be solved by the reallocation of resources within the discipline. These have already
been considered by other groups, and the Solar Physics Division
of the American Astronomical Society has established a committee
to develop specific recommendations. The present Working Group
has reached four major, unanimous conclusions with regard to
these problems.

First, the Working Group is greatly alarmed by the rarity
of solar-research programs at U.S. universities. This circum-
stance is due in part, we believe, to the mission orientation
of the funding agencies that supported solar physics from the
1950's to the early 1970's. This funding pattern has now
changed to some extent, with a larger portion of funding at
present coming from agencies that support basic research; however, the established institutional pattern of solar research,
which de-emphasized university-scale efforts, has not been significantly altered. The present imbalance in the distribution
of solar-physics research groups should be corrected, and we
believe that positive steps on the part of the funding agencies
(NASA and NSF) will be necessary to achieve a new balance.

1. We strongly support a temporary program of competitive
awards to young astronomers to encourage placing the best young
researchers in tenure-track positions at universities. We
believe that additional steps, specifically aimed at addressing
the present imbalance arising from lack of representation of
solar astronomers on university faculties, are also warranted;
however, a program benefiting astronomy as a whole is certainly
a positive first step, from which solar physics should itself
benefit.

2. The number, size, and quality of university solar
observatories (both optical and radio) has suffered severe
attrition in the past five years, and we urge that this
essential component of the facilities available for solar
research receive adequate funding in the 1980's. An expanded
program for instrumentation, detector development, and core
support for university observatories within the NSF grants
program for astronomy generally should have a positive effect
on solar-physics research in particular.

3. The continued vitality of the National Astronomy
Centers is essential for progress in solar physics. We are
particularly concerned about the failure to implement fully the
staffing and budget recommendations of the NSF Special Committee
for SPO, and the attrition in the solar experimental staff at
KPNO. The Centers must receive sufficient funds and must have
scientific staff adequate to carry out their missions.

Despite the unique perspective that radio observations of the Sun can provide, as demonstrated by their impact on a wide range of solar problems, solar radio astronomy in the United States has been neglected for more than 20 years. It will be important to increase the amount of grant funds available for the support of existing solar radio facilities, such as Clark Lake, Fort Davis, and the Sagamore Hill Observatory, and to support solar research at more general facilities such as Owens Valley. We also encourage NASA to be alert to opportunities to develop spacecraft radio instrumentation, which can make unique contributions to solar and heliospheric physics, through the Shuttle-science program.

4. We are concerned with the persistently low level of support that solar radio astronomy has received in the last 10 years and urge the following steps:

(a) The National Radio Astronomy Observatory should initiate a modest solar program by adding one or two solar radio astronomers to its staff; and

(b) Increased grant support for university-based research in solar radio astronomy should be made available.

VIII.   ACKNOWLEDGMENTS

The Working Group on Solar Physics is grateful for the advice and support of many people during the nearly 2-year-long effort required to complete this chapter. We wish to thank in particular the following: David Bohlin and George Newton of the National Aeronautics and Space Administration; Denis Peacock of the National Science Foundation; Paul Blanchard, Executive Secretary of the Astronomy Survey Committee; and our colleague Gordon Newkirk of the High Altitude Observatory, who chaired the Solar Cycle and Dynamics Mission Study Panel of NASA and whose report was most helpful.

2

Planetary Science

I. INTRODUCTION AND SUMMARY

The study of planets, which has been a basic part of astronomy
since antiquity, has undergone a significant transformation as
a result of modern technology. Unshackled in the 1960's from
the limitations of purely Earth-based observations, planetary
astronomy is now part of a new alignment of scientific interests
called planetary science.

In this chapter we attempt to trace the successes and weaknesses of this new field during its growth through the 1970's
and to outline the requirements that we believe need to be met
for substantial progress in the coming decade. In reporting to
the Astronomy Survey Committee, we were not asked to evaluate
opportunities for individual planetary deep-space missions, and
this has led to some difficulty in outlining programmatic opportunities. Such missions, and their careful coordination
with parallel efforts in the subdisciplines of ground-based and
Earth-orbital planetary astronomy, meteoritics, cosmochemistry,
planetology, and exobiology, are, we believe, at the heart of
future progress in planetary science and must be carefully
integrated into any program of the highest priority. As a
modus vivendi, this chapter incorporates not only our view of
the ground-based and Earth-orbital needs of planetary science
but also the recently developed strategy (Report on Space
Science 1975 (Space Science Board, National Academy of Sciences,
Washington, D.C. 1976); Strategy for Exploration of the Inner
Planets: 1977-1978 (COMPLEX, Space Science Board, National
Academy of Sciences, Washington, D.C., 1978); Strategy for the
Exploration of Primitive Solar-System Bodies--Asteroids,

Comets, and Meteoroids: 1980-1990 (COMPLEX, Space Science
Board, 1980) for in situ exploration of solar-system objects
formulated by the Committee on Planetary and Lunar Exploration
(COMPLEX) of the Space Science Board. If a comprehensive
scheme of priorities is to be obtained, we believe that the
separate development of a specific program for planetary
science--one that fully integrates the practical potentials of
ground, Earth-orbital, and specific planetary missions through
the year 2000--will be required.

This chapter was generated as a result of iterative review
by the Working Group members through the mail and by means of a
single public discussion at the 11th annual (1979) meeting of
the Division for Planetary Sciences of the American Astronomical Society in St. Louis, Missouri, which was designed to
provide a measure of direct community involvement.

Our report highlights what seems to us to be the remarkable
number of unexpected discoveries made through ground-based research as well as from planetary missions. In fact, we believe
that contributions made by ground-based research expose a current vitality in this field that has not been widely appreciated even in the planetary-science community itself. A second
theme that we emphasize is the future potential of Earth-orbital spectroscopy in planetary science, and we conclude that a
more detailed consideration should be given to the concept of a
dedicated Earth-orbital telescope for solar-system investigations.

We identify the two most clearly recognizable goals in
planetary science as being a strong drive to establish a broad
empirical basis for deciphering the cosmogony of the solar system and a continuation of ongoing efforts to attain a balanced
and more complete exploration and theoretical interpretation of
the current state of the planets and their satellites.

We envisage the first of these goals to be accomplished by
a combination of research efforts that include in situ observations of specific cometary nuclei and asteroids from rendezvous spacecraft; laboratory cosmochemistry of meteorites and
interplanetary dust; ground-based and Earth-orbital telescopic
observations of outer-planet satellite systems, cometary nuclei, asteroids, and other minor planets; the search for, and
the establishment of statistics for, extrasolar planets; theoretical and computer modeling; and astrophysical studies of
physical conditions in dense interstellar clouds, cloud
fragmentation, and protostellar objects.

The second goal requires the continued and aggressive
application of all the research techniques that are currently
used in planetary science, especially planetary deep-space
missions. Specific requirements that we single out as being
essential are new starts on planetary missions; increased
support to individual research projects that are not necessarily directly associated with space missions; access to existing

and planned ground-based and Earth-orbital facilities (including the Very Large Array radio telescope, current radar systems, large optical astronomical telescopes of all types, Space Telescope, and orbiting infrared telescopes); increased emphasis on laboratory astrophysics, including molecular and atomic spectroscopy, reflectance spectroscopy, and adequate parameterization of physical and chemical processes important to planetary science; access to advanced computers; a new emphasis on a search for extrasolar planetary systems; consideration of an Earth-orbital telescope essentially dedicated to solar-system work; and finally, the establishment of an informal occultation network.

II. THE NATURE OF PLANETARY SCIENCE IN 1980

A. The Planetary-Science Community and Its Goals

The planetary-science community is a phenomenon that has developed over the past two decades mostly as a result of major U.S. and Soviet Government programs to explore the solar system by spacecraft. It is a loosely organized, interdisciplinary community of scientists whose backgrounds, training, and interests are drawn from such diverse but more firmly established fields as aeronomy, astronomy, cosmochemistry, geology, geophysics, meteoritics, meteorology, physics, and space physics. We estimate that there are more than 700 scientists active in planetary science in the United States at present.

The primary scientific goals of this community may be most simply expressed as these:

1. The acquisition of knowledge about the current physical, dynamical, and chemical state of phenomena that occur in the solar system (other than the Sun itself);
2. The establishment of knowledge and understanding regarding the origin and subsequent evolution of the solar system and of its individual components; and
3. The achievement of sound explanations of phenomena that naturally occur in the solar system in terms of basic physical and chemical principles.

Underlying these broad goals are two further interests that deserve mention: the origin of life and a deeper understanding of natural processes that occur in the Earth, its atmosphere, and its oceans. With respect to the first of these, post-Viking interests appear to be centered on the chemistry ("prebiotic") of carbon in the primitive solar system as exemplified by small bodies (comets, asteroids, and meteorites) and perhaps also by the atmospheres of Titan and Jupiter. With respect to the second topic, the concept of "comparative planetology" has been put forward. In this concept the solar system is viewed as a giant laboratory in which many of the natural phenomena that occur on Earth are recognized to occur in other planetary settings but under diverse physical conditions and often at different evolutionary stages. It is anticipated, but only beginning to be demonstrated, that broadening our experience to such a divergent range of conditions will lead to a better understanding of the behavior of terrestrial phenomena.

B. Research Techniques in Planetary Science

While it seems clear to us that a major stimulus for the rapid growth of planetary science in the past decade has been the highly successful and aggressive program of in situ and remote observations from planetary spacecraft and probes, the advances in knowledge that have been achieved are, in fact, due to the application of a much broader range of research techniques. These techniques include observations from planetary spacecraft and probes; analysis of material samples returned from the Moon; all types of ground-based telescopic and radar observations (including observations from aircraft, balloons, and sounding rockets); elemental, isotopic, and mineralogical analysis of meteoritic material in cosmochemical laboratories; Earth-orbital spectroscopy; laboratory spectroscopy and parameterization of physical processes; and computer and theoretical modeling of complex physical scenarios.

While we do not expect any major additions or deletions to this list in the coming decade, we do expect considerable changes of emphasis. In particular, we expect that planetary scientists will become considerably more active in remote sensing from Earth-orbit as a result of the powerful instrumental potential of Space Telescope (ST), Infrared Astronomy Satellite (IRAS), and Shuttle Infrared Telescope Facility (SIRTF); we also anticipate an increasing emphasis on cosmochemistry as a result of new meteorite finds and newly developed methods for the acquisition, handling, and chemical analysis of exceedingly minute interplanetary dust samples. Computer and theoretical modeling and interpretive data analysis are expected to receive more emphasis in the 1980's partly because more capable machines are expected to become more accessible to planetary scientists and partly because of the large and growing data base on solar-system phenomena that is becoming available.

C. Funding in Planetary Science

The primary source of funding support for the planetary sciences is the Office of Space Science of the U.S. National

Aeronautics and Space Administration (NASA). Further support, either provided directly to planetary research programs (roughly $0.5 million/year), or more often and substantially to projects in closely related fields such as astronomy and earth sciences, is furnished by the National Science Foundation (NSF). During the past decade, the annual expenditures in the planetary program at NASA have been dominated by the requirements of the Viking Mars Mission and ranged from a high of $665 million/year in 1974 to a low of $170 million/year in 1978. This large variation in yearly funding is a reflection of the fact that most of these funds were directly associated with the hardware phases of specific missions that tended to be widely spaced in time. However, a base level of approximately $25 million/year has been regularly invested in data analysis, supporting research, instrumentation development, and advanced mission planning.

For the 1980's, only the sense of the pressures on funding levels seem clear to us. Strong advocacy seems assured for many of the increasingly complex and extended planetary-science missions now being contemplated in the NASA planetary program, and the recent paucity of new starts can be expected to amplify this advocacy. The substantial base of high-quality data on solar-system phenomena that has been derived (and not yet fully utilized) as a result of Apollo Lunar-Orbiter, Mariner, Viking, Pioneer, and Voyager missions, and as a result of a strong program of ground-based and Earth-orbital observations, can be expected to continue to expand rapidly and become increasingly accessible. This growth seems assured not only as a result of the already approved encounters with Saturn, Uranus, and Jupiter that will occur in the 1980's but also because of the prospects for major improvements in ground-based and Earth-orbital capabilities and techniques. We anticipate that these factors will lead to strong pressure for increased and wider funding of specific theoretical, interpretive, and observational research programs to individuals, as well as for a stronger program of advanced instrument development.

D. Planning in Planetary Science

The responsibility for planning--both for identifying research directions and for establishing a national strategy and priorities for solar-system exploration--has been largely shared by the Space Science Board of the National Research Council and NASA's Office of Space Science. Much of the rationale for the program in planetary science during the past decade can be traced to efforts of the Space Science Board in the late 1960's and early 1970's (Planetary Astronomy: An Appraisal of Ground-Based Opportunities, 1968; Planetary Exploration 1968-1975; The Outer Solar System, A Program for

Exploration, 1969; Opportunities and Choices in Space Science, 1974). More recently, a general strategy for the near future has been formulated by COMPLEX in Report on Space Science 1975 (1976); Strategy for Exploration of the Inner Planets: 1977-1987 (1978); and Strategy for the Exploration of Primitive Solar System Bodies--Asteroids, Comets, and Meteoroids: 1980-1990 (1980).

In parallel with these efforts, NASA has played a major role in highlighting specific research topics and mission opportunities through the encouragement of wide community involvement in ad hoc science working groups or advisory groups and a number of highly successful scientific workshops. These have provided an important stimulus to research through publication of transcripts or groups of research papers.

Planetary mission opportunities for the 1980's that, in our opinion, are well understood in terms of planning include Venus Orbiting Imaging Radar (VOIR); an International Comet Mission that takes advantage of the 1986 apparition of Comet Halley; Lunar Polar Orbiter (LPO); and a Saturn Orbiter/Probe mission (SOP^2). Galileo, a combined orbiter and probe mission to Jupiter, is already approved and being developed for a 1984 launch. We anticipate that future planning will be concentrated in two areas:

1. Establishing priorities and preparing for specific mission opportunities in the early 1990's, when a number of excellent Jupiter swingby opportunities to the outer planets occur. We also anticipate considerable community stress on the further geophysical exploration of Mars and the Moon as a result of the absence of any new mission starts since the Apollo and Viking landings and the existence of firm objectives and an active community interest. Finally, preparation for extended missions to comets and asteroids will command a high priority.

2. As a result of a growing appreciation of the potential for remote sensing of planetary phenomena, both from Earth orbit and from the ground, we also anticipate that future planning will emphasize a full integration of these potentials into the overall assessment of priorities. As we will discuss below, it is already possible to see a substantial rationale for an Earth-orbital telescope that is essentially dedicated to planetary-science objectives. However, in spite of its relatively low cost (as compared with that for planetary missions), we believe it is essential that the priority of such a project be established in the context of a complete program of planetary science that includes planetary missions.

E.   Research Activity and Reporting in Planetary Science

In order to illustrate (at least approximately) the trends in
research activity over the last decade, we have compiled
statistics on the production of research papers in the following major areas: lunar science, inner planets, outer planets,
satellites and rings, and the study of primitive objects
(principally comets, asteroids, and meteorites). These
statistics, which are based on citations in Astronomy and
Astrophysics Abstracts and on entries in the 1979 Hawaii
Bibliography on Planetary Satellites and Rings (Institute for
Astronomy, University of Hawaii), are illustrated in Figures
2.1-2.3; they demonstrate a strong growth of research in
planetary topics and a marked leveling off of activity in lunar
science after the conclusion of the Apollo program. The
sensitivity of the yearly production of research papers to the
successful conclusion of a planetary mission is also apparent,
particularly in the cases of lunar science and of research
relating to the inner planets.
     In the case of the outer planets, the impact of the Pioneer
flybys of Jupiter is noticeable, but the decade-long growth of
research activity in this field is primarily a result of
ground-based research. The impact of the 1979 Voyager flybys
of Jupiter is not included in Figure 2.2 but will presumably be
substantial. The research area that reflects the most frenetic
activity during the 1970's concerns satellite and ring systems,
which again illustrates the vitality that has been present in
ground-based research. It is surprising that the growth of
this subject does not seem to have been directly related to the
two major unexpected discoveries (the Io sodium torus and
Uranus' rings) in the field.
     There is an indication in Figures 2.1 and 2.2 of a
significant growth in the population of the planetary-science
community throughout the decade. In Figure 2.1, for example,
the yearly total of abstracts is seen to increase by
approximatey 60 percent over the decade, while the number of
authors participating in the annual meeting of the Division for
Planetary Sciences of the American Astronomical Society (Figure
2.2) has doubled in the past 5 years. A further interesting
fact is the relatively even distribution of research production
among the four major areas.
     We offer the following interpretation of the trends
illustrated in Figures 2.1-2.3 as a prognosis for developments
in the coming decade: the planetary-science community is
growing and can be expected to continue to grow for some time.
This can be expected to produce both increased pressure on
funding sources and increased competition for research
funding. The demonstrated capabilities of ground-based and
Earth-orbital research techniques, which have been responsible
for most of the activity seen in research on objects in the

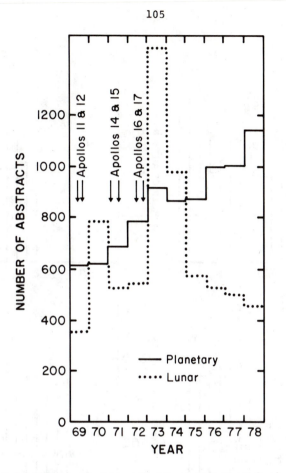

FIGURE 2.1 Number of research papers per year during the 1970's in the fields of planetary and lunar science.

outer solar system and on primitive bodies, will continue to be an essential part of planetary research for the foreseeable future. And, finally, if substantial further progress is to be made on scientific problems relating to the inner planets and Moon, then planetary space missions to them will be essential.

F. Education in Planetary Science

As a result of the interdisciplinary nature of planetary science, the training of planetary scientists takes place in many kinds of academic departments. In one planetary-science group for instance (at the University of Hawaii), the four

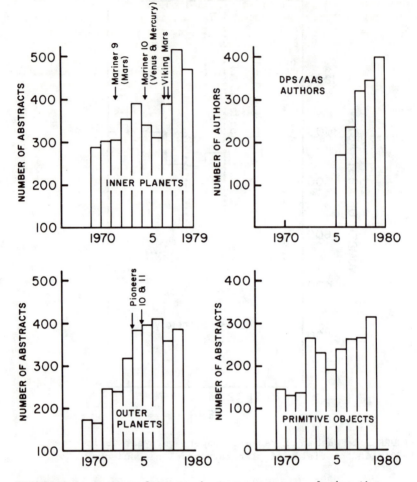

FIGURE 2.2 Number of research papers per year during the
1970's on the inner planets, the outer planets, and primitive
solar-system objects, together with the numbers of authors
participating in the annual meeting of the Division for
Planetary Sciences (DPS) of the American Astronomical Society
(AAS).

faculty members hold Ph.D. degrees in physics, astronomy,
geology, and chemistry. One of the strengths of the field lies
in the variety of disciplines that can be brought to bear on
research problems; however, a corresponding penalty is felt in
the absence of a common background or even a generally accepted
definition of planetary science. In general, the planetary

scientist is a minority member of a traditional academic department, such as astronomy, physics, or geology. In very few universities is the training of planetary scientists in the hands of a faculty specifically devoted to the discipline.

Consider three specific examples of universities that have recently granted a number of Ph.D.'s to planetary scientists. At Cornell, graduate training takes place formally within the Department of Astronomy, but in recent experience the majority of students have been in geology and, after completing their nonthesis requirements in the Department of Geology, these students have carried out their research with astronomy faculty as dissertation advisors. At the California Institute of Technology, planetary students are trained in the Division of Geological and Planetary Sciences, which is entirely distinct from the astronomy faculty; however, it is possible for planetary students to carry out observational work under certain

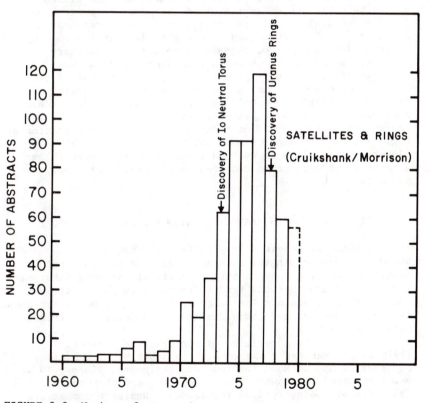

FIGURE 2.3 Number of research papers per year during the 1970's in the field of planetary satellites and rings.

circumstances with the Mt. Palomar facilities. A third mode of graduate education is represented by the University of Arizona, which has pioneered the establishment of an independent Department of Planetary Science, with faculty from a variety of disciplines including astronomy, meteoritics, geology, geoscience, and space physics. Observing facilities at Arizona are operated jointly by the Astronomy and Planetary Science Departments.

No accurate statistics exist on the number of Ph.D. degrees being granted each year in planetary science, but we estimate 15 to 20. The majority of these are in geology or related geosciences; not more than five of the recipients would be classed as planetary astronomers. During the past 5 years, most planetary-science theses have dealt primarily with spacecraft data or with the analysis of lunar samples or of meteorites.

In undergraduate education, it is a matter of some concern that most planetary science is taught by faculty with little or no training in this discipline. A large number of students study the solar system as a part of introductory survey courses in astronomy, almost all taught by stellar or extragalactic astronomers. The texts used rarely include contributions by persons with training or research interests in planetary science. As a result, there is a tendency to emphasize superficial aspects of a purely descriptive sort rather than to discuss the basic scientific issues that dominate contemporary planetary research. We believe that as long as undergraduates continue to learn about planetary science through astronomy courses an effort should be made to improve the presentation of this material, perhaps by more active participation of planetary scientists in writing texts, preparing curriculum material, and presenting lectures in elementary astronomy courses.

III. PROGRESS IN PLANETARY SCIENCE DURING THE 1970'S

The character of research in planetary science during the past decade has been dominated by the acquisition of fundamental data, discovery, and exploration. Rationalizations, interpretations, and explanations tend to be cautious and tentative. Questions proliferate. The solar system that is being revealed is one of contrast and individuality. Unexpected and strange phenomena abound.

About the time of writing of the Greenstein report (1970-1971), Mariner 9 was enroute to orbit Mars, the final sequence of Apollo Moon landings was beginning, Pioneer 10 was being prepared for launch toward Jupiter, and the Copernicus telescope was being readied for launch. Research emphasis was on the inner planets and the Moon. The Viking Mars project had begun, and there were cautious expectations regarding the presence of life on Mars; "Grand Tour" missions to the outer planets were being contemplated, and Jupiter was seen as the "Rosetta stone" of the solar system. In the intervening years, spacecraft have visited three planets (Mercury, Jupiter, and Saturn) and six satellites (Io, Europa, Ganymede, Callisto, Phobos, and Deimos) for the first time, successfully landed and operated on the surface of Mars and Venus, directly probed (simultaneously in five locations followed by a year or more in orbit in one case) the atmosphere of Venus, and returned material samples from several locations on the Moon. Important trends that have developed include the emergence of planetary satellite systems as a major area of study, a new emphasis on the study of the so-called "primitive" bodies (comets, asteroids, interplanetary dust, meteorites, and ring systems), and a growing awareness of the potential of Earth-orbiting telescopes and the complementary nature of Earth- and space-based observations.

Underlying the growth of knowledge has been the exploitation of new technology and the development of new research techniques and facilities. For example, the successful exploitation of Fourier-transform spectroscopy has tripled the number of molecules detected in the atmospheres of Jupiter and Saturn and has improved our knowledge of atmospheric structures on Mars and Jupiter. Improvements in infrared detectors and spectrometers and their application at airborne and large ground-based telescopes have opened up the discussion of satellite and planetary surface chemistry and of planetary stratospheres. The introduction of GaAs photomultiplier tubes and a variety of array detectors (reticons, microchannel plates, photoresistive strips, intensified vidicons, charge-injection devices, charge-coupled devices) has had an enormous influence on the performance of ultraviolet, optical, and near-infrared spectrometers for measuring weak emission or absorption lines in planetary atmospheres. The improvement of radar systems--including the resurfacing of the Arecibo antenna to work at S-band--has led to our first comprehensive views of the surface of Venus, a clarification of the nature of Saturn's rings, and further puzzles about the Galilean satellites of Jupiter. Heterodyne, Fabry-Perot, and very-high-dispersion grating spectroscopy using new fine-grain photographic emulsions is yielding detailed new information about global-scale mass motions in the stratosphere of Venus with a precision approaching a few tens of meters/second.

New techniques for handling and performing elemental and isotopic analysis of meteorites and microscopic samples of interplanetary dust have revealed unexpected results that have profoundly influenced our concepts of the primitive solar nebula. The 3-Meter Infrared Telescope Facility on Mauna Kea was completed by NASA in time to provide excellent 5-μm maps of Jupiter that are a vital complement to data on atmospheric phenomena obtained by Voyager, and the telescope is now in full operation. Interactive computer systems are being exploited in the establishment of regional data centers that allow easy access and interpretive processing of space mission data sets.

The Very Large Array (VLA) radio telescope is beginning to be applied to solar-system problems and has already provided a definitive result regarding the temperature and radius of Titan's surface. The VLA shows promise in its application to the Jovian radiation belts, and we have great expectations for its future applications to deep atmospheres and the surface of Venus. In space, the past decade has seen the introduction and successful deployment of the first U.S. planetary landers and atmospheric probes. Improvement of deep-space communication channels to incorporate both S- and X-band capability has increased the power of the radio-occultation technique to probe the vertical structure of atmospheres and allows high-rate real-time data transmission even for television.

There has been a steady increase in the size, weight, and capability of scientific payloads, together with improved stability and pointing of remote-sensing devices; finally, successful implementation of the planetary "swingby" and of multiple-encounter techniques has led to more frequent mission possibilities.

Scientific highlights include the discovery of ring systems around Uranus and Jupiter, which suggests that rings may be a much more common by-product of satellite-system formation than was previously thought. Further research into the dynamics and stability of these systems is expected to provide insight into the processes that form and dissipate such disks and that presumably occurred during solar-system formation. Three more satellites have been discovered, including Pluto's moon and the first member, Chiron, of a new class of small planets beyond Saturn.

The past decade has also seen the discovery of a multiplicity of asteroid surface compositions and evidence of systematic trends with heliocentric distance. The reflection spectrum of the most populous class of asteroids shows a similarity to that of carbonaceous chondrites, suggesting a generic relationship. All of these discoveries are thought to contain important information that will constrain our ideas of primitive conditions and dynamical rearrangements in the early solar nebula.

A particular surprise was the discovery of excess ^{26}Mg in almost pure alumina crystals in the Allende meteorite, indicating a high primordial ^{26}Al abundance and, hence, a large radioactive heat deposition in early solar-system materials. This and the measurement of other isotopic patterns in meteorites and samples of interplanetary dust are having a substantial effect on our perceptions of the range of physical conditions in the primitive solar nebula and the way the solar system could have been formed.

The detection and identification of the first "parent" molecules in comets was one of the main results of a well-coordinated, NASA-sponsored program to observe Comet Kohoutek in 1973. The detection of HCN, CH_3CH, and H_2O^+ points to intriguing similarities between cometary composition and the composition of interstellar molecular clouds. This connection, if true, identifies an important interface with astrophysics that clearly needs developing if we wish to understand the nature of possible presolar nebulae.

The decade has seen the discovery of magnetism on Mercury, Saturn, and possibly Uranus and exceedingly low limits put on Martian and Venusian fields. With the exception of the field of Uranus, all the measurable fields have been tentatively explored, but while many characteristics of their associated magnetospheres have been revealed, the details of the generation of these fields remain a puzzle. This is particularly so in the case of Saturn, whose field axis is observed to be closely aligned to the spin axis.

A huge advance in our knowledge of the Jovian magnetosphere and its interactions with the planet and satellites that are embedded in it has occurred. This is a result of a combination of data from the Pioneer and Voyager missions together with ground-based and Earth-orbital observations of Io-related "torus" emissions and Jovian auroras. Jupiter's magnetosphere is found to be dominated by heavy ions that are presumably derived from gases (including at least SO_2) observed to be released by spectacular volcanic activity on Io. The physical relationships between magnetospheric properties, Io volcanism, the latter's dependence on tidal stressing, decametric radio bursts, and planetary auroras have been partially worked out; a fascinating picture of highly complicated interactions is emerging that is beginning to explain many previously puzzling observations.

Aeronomy, like magnetospheric science, is not a major theme of this chapter because it is not highly susceptible to remote sensing. Nevertheless, a few major accomplishments should be listed. Both Mars and Venus have been directly probed by entry instruments, and Venus has been explored daily for more than a year. These atmospheres and ionospheres, while mainly analogous to the Earth's, show some puzzling differences. Most striking are the low exospheric temperatures, nowhere above 350 K and as cold as 100 K on the night side of Venus. The stratospheres of both planets, and even the near-surface atmosphere of Mars, display ozone chemistry closely analogous to what would be expected on a highly polluted Earth; no detectable ozone at all exists on Venus. Jupiter, probed only by radio occultations and ultraviolet spectroscopy, shows an opposite behavior, with very high exospheric temperature and other phenomena suggestive of a strong influence of magnetospheric plasma. In fact, the whole Jovian plasma torus can be regarded as an extension of Io's ionosphere, with suspicions of analogous but milder effects from Titan at Saturn.

The emergence of planetary satellite systems as a major area of study in planetary science was another important development of the 1970's. Increasingly sophisticated ground-based and airborne observations, together with the flybys of Pioneers 10 and 11 and Voyagers 1 and 2 through the Jupiter and Saturn systems, have opened up new fields of study concerning the large satellites of the outer solar system. In particular, the range of global properties and the surprising diversity of the surface properties of the Galilean satellites provide a new perspective on the evolution of small bodies and has given us our first examples of planetary-sized bodies for which ice plays a major role in both the surfaces and interiors of the objects.

A theory has also been developed that explains the origin and evolution of the orbital resonances among these satellites with dissipation in the interior of Io as the controlling factor. Bounds on the Q of Jupiter have been estimated and the negligible amplitude of libration of the three-body Laplace relation now has a reasonable explanation. This new understanding of the Laplace resonance formation is a fitting climax to the considerable development of the theories of origin of orbital resonances during the last decade.

The remarkable individuality of the surface properties of each Galilean satellite exposed by Voyager is not apparently confined to the Jupiter system. Excellent Viking images of Phobos and Deimos have shown an equal diversity of that system. These satellites are believed to be asteroidal debris captured by Mars during its formative stages, and Viking may have given us our first look with high spatial resolution at asteroidal objects.

The Earth's Moon continues to be an enigma. Absolute ages and rock types are now available for major geologic features on the Moon and, with the addition of heat-flow, seismic, and magnetic data, we now have a broad picture of its overall physical state. It is an ancient, highly differentiated body and perhaps partly molten in its interior. Its elemental composition is found to be different from that of the Earth, being deficient in the more volatile elements, while certain isotopic patterns (e.g., $^{18}O/^{16}O$) are similar to terrestrial patterns. This latter similarity, in contrast to the range of isotopic ratios seen in meteorites, places a significant constraint on lunar origin. While it still is not possible to choose confidently among all the major competing theories of its mode of origin, those postulating a distant origin may be taken far less seriously.

Controversy over the surface temperature of Titan seems to have waned as a result of recent measurements made with the VLA that clearly separate the radiation of the satellite from that of its parent planet. Nevertheless, much still remains to be learned regarding the extent, composition, photochemistry, and evolution of Titan's atmosphere; further analyses of the data from the Voyager flybys through the Saturn system should help to clarify some of these issues. The view of Titan as subject to a major greenhouse warming comparable with that of Venus, has, however, been abandoned.

With respect to the planets themselves, it is their individuality and often the unexpectedness of discovery that seems to stand out from the research of the last decade. A few maddening problems, such as the composition of the visible clouds of Venus and the general composition of the Martian and Venus atmospheres, have been solved. But the plain fact is that many more problems have presented themselves as a result of the last 10 years of successful exploration and unabated surprise.

At Mercury, which was found to be devoid of an indigenous atmosphere, the fraction of the surface that we have seen shows the record of what was presumably the same ancient heavy bombardment that is evidenced by the surfaces of the Moon, Mars, and Callisto. Signs of evolution in its interior show up in Mercury's own special brand of surface features: widespread volcanic plains--quite different from those on the Moon--and planetary-scale scarps that suggest that internal shrinkage has occurred. The major surprise of the Mariner 10 mission at Mercury was the discovery that this slowly rotating planet possesses a dipole magnetic field; this fact can be expected to be a significant datum for future explanations of planetary magnetism.

At Venus we have had our first glimpse of the surface terrain, both by radar and by direct Soviet television link. The natural radioactivity of surface rocks has been measured in three locations, being found to have similarities to several terrestrial igneous rock types, and it now seems probable that Venus is (or was) a tectonically active planet. The ground-based radar coverage already reveals a rugged surface with mountain ranges, possible volcanoes, chasms, plateaus but no indications yet of Earth-like crustal plates or constructs similar to terrestrial continents. There are also indications of craters that could be of impact origin. All of this is enough for us to urge efforts for a proper geophysical exploration of the suface of Venus in the future.

We can also report significant advances in our understanding of the composition, structure, and dynamics of the atmosphere and clouds on Venus as a result of the Pioneer Venus mission and much ground-based research. The omnipresent upper clouds are now known to be composed of H_2SO_4 (thanks to a theoretical analysis of ground-based optical polarization data) and to be controlled by the photochemistry of SO_2. Lower down in the atmosphere, other cloud layers involving H_2O and some as yet unidentified dark components have been charted. The primary constituents of the neutral atmosphere are now known with precision as a result of in situ mass spectroscopy and gas chromatography. A major surprise was the unexpected discovery of a relatively large abundance of ^{36}A in the atmosphere (20-100 times as abundant per gram of rock as on Earth). This finding is causing a major reassessment of solar nebula condensation scenarios for planetary formation.

In the upper atmosphere, a very complex (and still poorly understood) global structure has been observed in which a dynamic ionosphere, continually buffeted by the solar wind, is embedded. The 4-day "apparent" rotation of the lower strato-sphere/upper tropospheric regions, which is some 60 times more rapid than that of the planet itself, is now known to be a true mass motion of the atmosphere, although large-scale wave motions are thought to be embedded in it, and the vertical shear of zonal motions in the lower atmosphere has been measured. The curious Y-shaped markings seen since the 1930's in ultraviolet photographs appear to be associated with a global system of large-amplitude waves, but the mechanisms that produce these phenomena are still unknown. Important data regarding the radiation balance within the atmosphere have been obtained, and it is now known that appreciable fluxes of sunlight penetrate down to the surface; however, an entirely satisfactory explana-tion of how the high surface temperature (755 K) is maintained has not yet been worked out.

At Mars perhaps the most significant discovery was the effectively total absence of organic material on the surface. As a result, the search for life on that planet appears to have come at least temporarily to a halt, although one must recall that only two locations were sampled. The Viking mission and the Mariner 9 orbiter preceding it have, however, added much to our knowledge of the way in which planets and their atmospheres evolve. There is clear evidence on the surface of Mars that volcanic and other tectonic events occurred early in the planet's history; yet, wide expanses of the planet also show primitive cratered landscapes untouched except for atmospheric erosion. Some of these craters have forms that suggest that a large reservoir of water lies beneath much of the surface in the form of permafrost. Also found on the surface are relics of what are almost certainly stream and flood channels. Cer-tain types of these channels suggest that there may have been a period in the early history of Mars when its climate was far more clement than now. A second indication of changes in the climate on Mars is the presence of layered structures in the terrain at the edges of the polar caps. These changes are likely to be the result of long-term variations in the planet's orbit about the Sun. The exact chronology of Martian atmospheric history is uncertain, but evidence for a denser atmosphere in earlier times has also been inferred from mea-surements of the relative abundances of nitrogen isotopes.

As a result of our newly acquired knowledge of the detailed composition of the Martian and Venus atmospheres (together with the composition of the Earth atmosphere), we have seen the beginnings of efforts to construct comprehensive theories of atmospheric evolution in the inner solar system. In addition, for Mars, precise data on atmospheric composition, transparency, structure, and motions (from wind streaking and meteorological stations on the surface) have led to an increased understanding of the nature of seasonal albedo variations that have long been observed from the Earth, the details of dust storms, and the overall mechanics of the atmospheric circulation (which turns out to have considerable similarities to that of the Earth).

In the realm of the giant planets, some of the most substantial advances have been in our knowledge of chemical abundances. Almost entirely as a result of the introduction of Fourier and Fabry-Perot spectrometers at ground and airborne observatories, our knowledge of atmospheric constituents has grown from a short list of four molecules (H_2, He, CH_4, and NH_3) to encompass C_2H_2, C_2H_6, CH_3D, HD, CO, GeH_4, H_2O, and PH_3 on Jupiter and CH_3D, C_2H_2, C_2H_6, PH_3, and HD on Saturn. Stratospheric emissions of C_2H_4, C_2H_6, and posssibly C_2H_2 or C_3H_8 and CH_3D have been detected on Titan and C_2H_6 on Neptune. HD has been detected on Uranus and Neptune. Knowledge of relative abundances are still very uncertain, but it does seem clear that at least in Jupiter's and Saturn's deep atmospheres there are substantial deviations from local chemical thermodynamic equilibrium and that vertical mixing of hot deeper regions with the cold visible atmosphere is an important process. As mentioned above, an accurate assessment of relative abundances still eludes us; however, major progress has been made for some constituents. Making use of a wide range of spectroscopic observations (including center-to-limb variations) and also theoretical analyses that couple the solution of the abundance problem to the problem of vertical atmospheric and cloud structure, we have learned that there exist very distinct compositional differences between the visible atmospheres of Jupiter and Saturn on the one hand and those of Uranus and Neptune on the other. This finding furnishes an important constraint on some of the details of the modes of origin of these planets.

Substantial advances have also occurred in our knowledge and understanding of the vertical structure of outer-planet atmospheres. These advances have resulted from a complex combination of ground-based and space-based observations. For example, ground-based infrared spectrophotometry has revealed that strong stratospheric thermal inversions are the general rule on the outer planets (although Uranus may be an exception); together with vertical sounding of this structure by means of ad hoc networks of ground and airborne observations of stellar occultations by Jupiter, Neptune, and Uranus, there has been stimulated much discussion of seasonal behavior (for Saturn), photochemistry, energetics, and transport mechanisms in stratospheric regions.

From space it has been possible to obtain precise knowledge of the structure of Jupiter both in its upper-atmospheric regions and in the troposphere. Unexpectedly, Jupiter's upper atmosphere was found to have a very high electron temperature (about 1000 K) and a much more extensive ionosphere than anticipated. In addition, widespread auroral activity, possibly related to field lines threading Io and Io's torus, has been found in hydrogen Lyman-alpha emission and band emission of H_2. These polar auroras have also been observed from the International Ultraviolet Explorer satellite, and we can expect Space Telescope to provide important information on these phenomena in the future. In the troposphere, thermal radiometry from Pioneers 10 and 11 found that meridional gradients in outgoing thermal fluxes are minimized by an apparently increased outflow of internal flux at higher latitudes, a factor that should be an important consideration in understanding the origin of the belt/zone structure.

An enormous but as yet undigested data base has been established for Jupiter concerning cloud structure and its correlation with color, atmospheric motions, and atmospheric electricity. A few preliminary results regarding the stability of zonal flows, their relationship to underlying turbulent motions and to the pattern of colors, are beginning to emerge, but a clear picture will require much more research. The striking Voyager television pictures of Jupiter, together with similar data for Mars and Venus, are expected to nourish and clarify theories of atmospheric dynamics for many years.

Progress can also be reported in our knowledge of the outermost giant planets, Uranus and Neptune. Perhaps the most striking new fact has been the unexpected discovery of an appreciable internal heat source within Neptune and its absence from Uranus. A possible explanation has been put forward in terms of the sensitivity of the outward transport of internal energy from the interior to the external boundary conditions imposed by solar radiation. Several new observations point to other possibly related differences in the atmospheres of these two planets. For example, there are strong indications of clouds and weather and a definite stratospheric inversion on Neptune, but such phenomena are apparently absent on Uranus. Other observations have revealed major weaknesses in our techniques or understanding. The microwave brightness of Uranus is currently unexplained except in terms of an unlikely absence of NH_3 deep in the atmosphere. This probably indicates that we may be surprised by its atmospheric structure, if not by its chemistry. Also, numerous attempts to measure the rotation periods of these two planets, using the best ground-based equipment and modern techniques, give widely divergent results, an embarrassing situation for remote sensing that needs to be understood.

IV. MAJOR OBJECTIVES IN PLANETARY SCIENCE FOR THE 1980'S

A. Cautionary Statement

In a report such as this, there is always a danger of placing inadvertent constraints on the development of a subject, either by overspecifying the nature of a problem or by ignorance or simply by inadvertent omission. These dangers are particularly acute if the users of the document (who are presumably those primarily resonsible for funding research) interpret too literally what is written. Thus, as an introduction to this and the following section, we emphasize that what is presented here is simply our point of view at this specific time; given the high frequency of discoveries, the rapid changes in tech- nology, and the massive amount of new data still being har- vested by Pioneer Venus, Voyager, and Viking, plus the very recent successful flyby of Pioneer Saturn, it is possible that our point of view may undergo significant changes within short time scales. In other words, the rate of advance of new knowledge is sufficiently fast to make any priorities that we established today valid only over very short times--times that are certainly short when compared with a decade.

B. Some Outstanding Problems in Planetary Science

The general scientific goals that were expressed in earlier serve to define the breadth of interest in planetary science. They do not, however, indicate the spirit of the research at any particular time. In this section we try to communicate this spirit in terms of a collection of specific problems that seem central to the work being done today and possible for the near future.

1. Origin of the Solar System

Specific problems of current interest that could (and in our opinion should) be resolved in the next decade are the following:

• To what extent are small bodies in the solar system (e.g., cometary nuclei, asteroids, meteorites, interplanetary dust particles) representative of unmetamorphosed primitive material?

A resolution of this question is important, for on the answer rests our ability directly to sample materials accreted at the time (and possibly before) the collapse of the interstellar cloud that formed the solar system. It is the educated assumption that some small solar-system objects are pristine, which makes them such important targets for future space missions and for ground-based and Earth-orbital observational programs.

• What physical relationships exist between the meteorites, interplanetary dust, and cometary nuclei and/or various classes of asteroids?

If direct physical relationships can be proven, then they would allow confident assignment of the physical conditions inferred from the meteoritic record to specific regions of interplanetary space.

Other questions in the same vein are the following:

• What are the chemical and physical characteristics of cometary nuclei and asteroids?

• What was the original physical state of the solar nebula? Do the types of nebulae envisioned in current cosmogonical theories actually exist in nature?

• What was the primitive chemical state of carbon compounds in the solar nebula?

• Are there other planetary systems, and what are their statistics?

• How complete is our current inventory of solar-system objects? Are there new classes of objects to be found?

It should be clear from the nature of these questions that they embody the real possibility of establishing an empirical basis for studying the cosmogony of the solar system. The recent and unexpected discovery of isotopic anomalies, indicating that presolar interstellar material is an available ingredient in meteorites; the detection, in a comet, of molecules that also exist in interstellar clouds; the recognition of distinct chemical classes of asteroids and the strong indication in them of chemical gradients as a function of distance from the Sun; the tentative association of the largest class of asteroids with carbonaceous chondrites--all of these recent discoveries argue to us that the smaller bodies in the solar system represent a rich treasure of information about the early solar system.

The means for capturing this information can be made available over the next decade. A comet rendezvous mission can make an in situ exploration of the physical structure of a comet nucleus, and its chemistry can be assessed with great precision by current mass spectrometers. The body of the nucleus and its structure can be probed by radar; its physical and chemical heterogeneity can be observed by a host of techniques as the comet disaggregates during its perihelion passage. A second important venture that would produce otherwise unavailable knowledge on primitive objects is a rendezvous mission to a selection of asteroids.

Ground-based radar systems are now capable of probing the general physical characteristics of the surfaces of about 60 known asteroids and possibly some cometary nuclei within the next few years. Much more remains to be done in the near-infrared region (1-5 μm) and in the ultraviolet below 0.3 μm to extend our knowledge of asteroid surface chemistry. Millimeter-wave spectroscopy of comets should now be sensitive enough to increase our knowledge of their molecular constituents, and predictable advances in molecular cloud astronomy will surely lead to much more information about cloud fragmentation, protostellar objects, and the possible range of conditions in putative solar nebulae.

Such research should also lead to a better definition of the conditions in our own primordial nebula in which planetary formation is known to have occurred. Current models of solar-system formation all assume reasonable initial conditions but are otherwise ad hoc; it may be that none approximates the truth. Three-dimensional hydrodynamic numerical codes constructed to follow the details of protostellar collapse may lead to a more realistic definition of these conditions, and the continued development of such codes should be encouraged.

Refinement of astrometric techniques should make it possible to push the limit of detectability of the masses of secondary objects in a large number of binary systems well below the current limits of approximately 0.1 solar mass. The detection of substantial number of 0.001 solar mass objects seems possible. As a result a respectable number of extrasolar planets of the mass of Jupiter might be found and thereby lead to a statistical basis for cosmogonic theory in much the same way that, during the past 20 years, photometry of a wide selection of stars established an empirical basis for the theory of stellar evolution.

Finally, it should be noted that our inventory of the various classes of small solar-system objects is probably not complete. The exciting discoveries of Pluto's moon, Chiron, and various satellites, rings, and Earth-crossing asteroids during the past 10 years attests to this. One of the important and difficult problems of the next decade is, in fact, to continue this exploration for new small and possibly distinct classes of objects. This program may yield important links between meteorites, impact flux, asteroids, and cometary nuclei. The discovery of new satellites (particularly if related to ring systems) should provide further insight into the dynamics and evolution of rings. A further benefit would be the compilation of a comprehensive catalog of small objects that would be good candidates for future space missions, either for reasons of technical performance or as representatives of special classes of objects.

2. Planetary Atmospheres

In this field, the spirit of research is increasingly turning toward problems of evolution and stability. For example, there is considerable interest in the nature of the climatological changes that appear to have occurred in the early history of the atmospheres on Mars and Venus. The presence of fluvial channels on Mars and the observed $^{15}N/^{14}N$ ratio seems to indicate an early period in which the Martian atmosphere was much more substantial than it is at present. The remarkable differences in atmospheric conditions, particularly the contrast in the abundance of water on the Earth and on Venus, have led to investigations into the stability of atmospheres with respect to small changes in solar insolation. Nevertheless, the primary emphasis at present still centers on the important work of extracting factual information from an observation (e.g., chemical abundances) or in the derivation of mechanistic explanations of specific phenomena. Current and future research topics include the following:

Questions of Dynamics

* What gives rise to the banded appearance of Jupiter and Saturn, and what is the physical relationship between the observed coloration and motions?
* What is the nature of the Great Red Spot and other spots on Jupiter?
* What is the cause of equatorial jets on the outer planets?
* How is the superrotation of the atmosphere of Venus maintained? What is the nature of the famous ultraviolet "Y" feature?
* What is the cause of the frequent dust storms on Mars?
* Why is the nightside "thermosphere" of Venus as cold as 100 K?

Questions on Composition and Evolution

* What are the relative elemental and isotopic abundances of H, He, C, N, and O on the outer planets, and how do these compare with the solar values?
* What constraints do these abundances place on our ideas regarding the processes involved in the formation of the outer planets and their subsequent evolution?
* How does photochemistry of species of H, C, N, S, O, P proceed in the upper layers of both inner and outer planetary

atmospheres? How does this effect atmospheric structure,
evolution, cloud or haze formation, and atmospheric coloration?

* What are the abundances of rare gases and their
isotopes on Jupiter and the outer planets, and how was the (now
observed) abundance patterns of these gases on Earth, Venus,
and Mars established?

* Can we utilize the presence of observable nonthermo-
dynamic equilibrium compounds on Jupiter as tracers of mixing
processes on Jupiter and Saturn?

* How have conditions in the Martian and Venusian
atmospheres evolved over geologic time scales? Was there once
water in the atmosphere of Venus? What was the early climate
on Mars?

* What role have cometary impacts had on the composition
of the atmospheres of Mars, Venus, and the Earth?

Questions on Structure

* What is the structure of the major cloud layers on the
outer planets, what form do these clouds take, and what role do
they play in the energetics of the visible atmosphere?

* What is the cause of the hot upper atmosphere on
Jupiter? How interrelated are the energetics of the upper
atmosphere and Jupiter's magnetosphere?

* How is the high temperature of the Venus surface
maintained?

* Why do Neptune and Uranus have such different
stratospheric structures?

* What are the primary physical processes that govern the
structure of outer planet stratospheres?

At present we are witnessing a tremendous growth in
empirical knowledge about motions in planetary atmospheres.
From space, Voyager images are mapping out the zonal flow and
distinguishing in a preliminary way the relationships between
organized flows, eddy fields, and possible wave propagation.
Similarly Mariner 10, Venera probes, and Pioneer Venus have
begun to chart the three-dimensional nature of the planetary
circulation of the lower atmosphere on Venus. On the ground,
high dispersion and resolution Doppler spectroscopy is
beginning to map out with increasing confidence the horizontal
flow fields that exist in and above the cloud tops in the
atmosphere of Venus.

We see beginning to emerge in these developments a solid
data base that should lead to confident, mechanistic explanations of the features that dominate the appearance of deep
atmospheres--belts, zones, spots, Y's--and also to a better
understanding of the overall energetics of planetary atmospheres.

In the next decade it is vital that this data base expand to encompass and delineate the relevant time and spatial scales of the phenomena of primary interest and also that the proper theoretical and computational tools are developed. We anticipate at least the following developments: The Galileo mission and Space Telescope (ST) will provide new insights into the motion fields on Jupiter, for, as a result of their ability to image Jupiter in the region of strong near-infrared absorptions, their camera systems should be able to probe the three-dimensional structure of the motion field. ST will be able to record the motions on the largest and longest time scales, while Galileo will be limited to very-high-resolution studies of the dynamical properties of specific features. The Galileo atmospheric probe should provide detailed ground truth for these measurements at the region of probe descent through measurements of structure and the local wind shear.

Ground-based observations using Doppler techniques can, if applied in a systematic way, yield much information about global properties of the motion fields in the stratosphere of Venus and help to distinguish between planetary-scale wave motions and the underlying flows. A new development that may be useful in the future is the tracking of motions in the deep atmospheres of Venus and Jupiter by following thermal anomalies with the VLA. Effective interpretation and explanations of these phenomena will require the extension of three-dimensional models to finer computational grids than are used at present, so that the great depth of the atmospheres of Venus and Jupiter can be properly simulated and yet have adequate zonal and meridional sampling.

With regard to composition and structure, the Galileo probe will provide precise knowledge on the elemental and isotopic composition of the Jovian atmosphere and its thermal structure down below the base of its visible atmosphere (about 10 bars). Together with what has been learned from Venus and Mars, this information will help to clarify not only the details of Jupiter's atmosphere but also our view of topics ranging from the chemistry of the solar nebula to the general structure of planetary interiors. These Galileo results will also provide important ground truth for remote sensing of composition and structure both from the ground and Earth orbit.

In remote sensing, the prime emphasis in the past has been on the detection of specific constituents, although with increasing efforts to measure the global morphology of spectral absorptions. This changing emphasis is a result of the fact that, as far as remotely sensed data are concerned, the structural and compositional problems must be solved together. We expect this trend to lead to an increased emphasis on "imaging" spectroscopy (in all spectral regions) both from the ground and from space. In this regard, the combination of high spectral and spatial resolution in the spectrographs being developed for

ST is of special importance, as is the future development of high-resolution infrared spectrographs for both Space Telescope and SIRTF. Already efforts to combine Fabry-Perot and Michelson spectrometers with array detectors or vidicons promise a powerful capability for imaging spectroscopy in the optical region, and this development should propagate into the infrared region as low-noise panoramic detectors become available for use on instruments on large telescopes. One obvious infrared application is the mapping of stratospheric emissions and of absorption lines of nonequilibrium constituents on Jupiter and Saturn and study of their temporal stability.

In the ultraviolet region, Earth-orbital data combining good spectral and spatial resolution can be expected to play an important role in elucidating and constraining our ideas of the nature of aeronomical processes in the upper atmospheres of Venus and Mars and also in helping to disentangle information about auroral and stratospheric phenomena in the outer planets.

3. Planetary and Satellite Surfaces

In order to effect major advances in this area the primary need seems clear: except for some level of exploratory work, detailed investigations by orbiting and landed spacecraft are essential. The Galilean satellites, for which preliminary morphological maps showing the general character of the surface are now available, should be adequately covered before the end of the decade from the point of view of both chemical and topographical mapping by the Galileo orbiter. For the inner planets new starts on orbiting spacecraft are obviously required, since only by such means can the necessary detail and coverage be obtained. A radar imager for Venus is feasible that can provide global coverage of the planet at 0.6 km per line pair resolution or better. Experience shows that this level of resolution is essential if the morphology of the surface is to be interpretable in terms of geologic processes. Such observations are not only our major hope for understanding the evolution of the interior of Venus but may also yield information on the development of its remarkable atmosphere and the reasons why it is so unlike that of the Earth.

In the cases of Mars and of the Moon, orbiting spacecraft that can investigate the chemical nature of the surface are urgently needed in order to obtain information to complement the high quality of topographic and morphological data already available and the chemical sampling done at specific landing sites. The Viking orbiter mission established a global view of Mars to resolutions better than 0.5 km/line pair and in some areas to resolutions of 16 m/line pair. What is still missing is spectral information that should allow a clear differentiation of mineralogical, chemical, and textural contrasts in the geologic constructs that have been found. Such information, together with other types of geophysical data on the surface, is necessary for reliable inferences regarding the volume and physical conditions in the planet's crust, its mantle, and its core. This information will also help to determine the history of surface material through volcanic, sedimentary, and weathering processes.

Questions of Primary Interest

* What are the major chemical components and regimes on the surfaces of Mars, the Moon, Venus, and Mercury, and how are they distributed relative to topography and morphology?
* What can the topography of the surface of Venus tell us about the processes that have occurred (and are possibly still occurring) in its interior? About the history of its atmosphere?
* What is the current surface heat flow on the inner planets?
* Why does only the Earth possess water oceans?
* Why have Mars, Venus, and the Earth evolved in such different ways?
* Why have the surfaces of Galilean satellites evolved so differently, or at such different rates?
* What are the energetics of the volcanism on Io?
* Do Europa, Ganymede, and Callisto still have active interiors?
* What is the surface of Titan like? Are there clues to the evolution of the unique atmosphere that exists on this satellite?
* What does the "other" side of Mercury look like? Does it show the global dichotomy so characteristic of the Moon, Mars, and the Earth?

Much of the research that we foresee in this area will depend, at least for the near future, on the data base derived from past orbiting spacecraft. Ground-based or Earth-orbiting systems do not provide high enough spatial resolution for this purpose. Nevertheless, there are a few exceptions: ST can, at a few favorable opportunities, yield a glimpse of the currently unseen side of Mercury at resolutions of about 60 km/line pair. This resolution is high enough to recognize any major physiographical provinces that exist, to see if there are any indications of gross global dichotomy of terrain, and to permit comparisons with the hemisphere observed by Mariner 10. Important opportunities for doing this (which unfortunately will present a difficult operational maneuver for the telescope) exist at aphelic elongation in 1984, 1985, and 1986, after which there will be a hiatus for several years.

ST should also be useful for some preliminary chemical mapping on Mars, where it can provide spatial resolution between 35 and 140 km/line pair. It may be possible by such observations to follow changes that provide insight into atmospheric-surface interactions. In a much cruder way ST can also acquire images of the Galilean satellites in the near ultraviolet; a spatial resolution of 250 km/line pair is possible with the wide-field planetary camera. It is difficult to predict what these bodies will look like in the UV, but there are strong indications from ground-based photometry that substantial, and therefore interesting, UV contrasts on these surfaces do exist.

Technical improvements to the Goldstone radar facility have been proposed that could lead to higher-quality topographic information on Mars, Venus, the Moon, and Mercury, although the general topographic contrasts in a global sense are reasonably well known at present. The nature and texture of the surfaces of the Galilean satellites, which have produced remarkably strong and unexpectedly polarized echos, clearly require further investigation. The detection of radar echos from Titan seems a distinct future possibility, and there is much interest in what they will reveal about Titan's surface.

The radar mapping of Venus from Arecibo was completed in 1980 and covers roughly one third of the surface at a resolution of about 20 km/line pair and about 10 percent of the surface at 8 km/line pair. Currently available radar images, together with the global topographic maps (at substantially lower spatial resolution) obtained from the Pioneer Venus orbiter, have presented us with an intriguing picture of the surface of this planet. There appears to be no obvious evidence for the kind of plate tectonics seen on Earth, but there are substantial indications of mountain building and other possible tectonic features. These features show small associated gravity anomalies, presumably indicating a plastic crust. In some areas there are indications of craters, but the spatial resolution is not currently enough at any of these features for a secure geologic interpretation. Our prospects for a deeper understanding of the surface of Venus depend primarily on future planetary space missions.

4. Planetary Magnetism and Magnetospheric Interaction

The last decade has seen many new and remarkable facts learned about planetary magnetism. The magnetic field of Mars (if it has one) is of unexpectedly low strength. The Moon has unexpectedly strong remnant magnetism. Surprisingly, Mercury possesses a relatively strong dipole field. The Saturn field appears to be aligned with its rotation axis, which, according to simple dynamo theories, should lead to its decay. Uranus almost certainly has a strong field, and Voyager, which should reach Uranus in 1986, can be expected to provide further information. By a process of elimination, the sources of planetary magnetic fields are thought to be internal dynamos involving the motions of electrically conducting fluid cores. Mercury's dynamical configuration affords a unique test for this hypothesis if the extent of a molten core can be determined. Whether Mercury's field is dependent on a molten interior is a piece of information vital for the development of theories of the origin of planetary magnetic fields.

Magnetospheric interactions in the outer solar system-- particularly the relationships between the Jovian magneto- sphere, Jupiter's upper atmosphere, Io's volcanism, and Io's surface--seem destined to be a very active field in planetary science for the foreseeable future. Its study can be effec- tively pursued with high-resolution and imaging spectroscopy from the ground and Earth-orbit (ST) and by in situ measurements with the Galileo spacecraft. The insights that we have already attained come from such complementary observations made from Voyager, IUE, Copernicus, and with ground-based telescopes. The phenomena associated with these interactions are already known to be highly dynamic, and it seems to us that their future understanding will require a comprehensive set of observations with extensive time and spatial coverage. As a result of the wide utility of ST and IUE in other areas of science, we anticipate that there will be difficulties in obtaining the necessary time coverage of ultraviolet phenomena associated with these interactions, which must be done from Earth orbit. In addition, there is a major deficiency in our ability to obtain future spectroscopic coverage for the region 500-100 Å. It is in this spectral region that the Voyager ultraviolet spectrometer found the bulk of the energy radiated from the Io plasma torus. These facts emphasize the growing need for consideration of a dedicated telescope in Earth orbit for solar-system observations, as discussed in the section on programmatic opportunities. Some of the leading questions related to planetary magnetism are the following:

* How strong is Uranus's field, and what is its relation to the planet's rotation and inner structure?
* Do the Galilean satellites, in particular Io, have magnetic fields?
* Is the current theory of dynamo action really relevant to planetary-field generation? How do we explain the con- figuration of Saturn's field?
* Does Mercury have the necessary molten core for dynamo action?

A selection of important questions relating to magnetospheric interactions follows:

* How and where are ions created and accelerated in the Jovian and Saturn magnetospheres?

* How stable is Io's plasma torus?
* What physical processes maintain the plasma torus and the strong temperature gradients within it?
* How are neutral sodium and other atoms ejected from Io?
* Where is the source region of Io-related decametric radio bursts located?
* How are Jovian auroras related to the Io torus? How does the interaction proceed?

5. Planetary and Satellite Interiors

Advances in the study of the structure of planetary interiors depend on a broad variety of inputs, including accurate data on global properties (gravitational harmonics, rotation period, figure, composition, mass, and heat loss) and accurate data on equations of state and phase transitions for naturally occurring mixtures. We understand that substantial progress is being made at present in the investigation of the latter as a result of the development of diamond cells, with which static measurements can be made at very high pressures. We strongly encourage further work in this area, especially for mixtures of planetological interest, including refractory materials, geophysical ices, and mixtures of H and He.

With regard to accurate knowledge of the global properties of satellites and the more distant planets there remains much to be learned. A recent application of ground-based speckle interferometry to the dimensions of Pluto looks encouraging, and we call for more research with this technique. Problems about which we expect wide discussion in the next decade include the following:

* To what degree has helium differentiated from hydrogen on Jupiter and Saturn?
* Why do Uranus and Neptune have different thermal luminosities?
* What are the rotation periods and figures of Uranus and Neptune?
* Are current models of the thermal histories of the Galilean satellites and Titan adequate?
* What was the thermal history of the asteroids and meteorite parent bodies?
* What are the energetics of Io's volcanism?
* Are current models of internal heat generation close to the truth?

Significant advances in these problems will again require data obtained from both Earth-orbital telescopes and planetary spacecraft. The rotation periods of Uranus and Neptune will probably not be known with an accuracy adequate to attack these problems until ST is available. High-quality data on the thermal spectrum of Uranus and Neptune in the 10-100-μm range will require observations from SIRTF. Voyager will improve on existing data of the thermal spectrum of Saturn. Accurate data on the thermal luminosity and energy balance of these planets will require observations from future spacecraft. Information on the lower gravitatonal harmonics of Titan and of one or two Galilean satellites will also become available as a result of the Voyager Titan flyby and the orbital satellite tour by Galileo.

V. PROGRAMMATIC OPPORTUNITIES FOR PLANETARY SCIENCE IN THE 1980'S

We expect that the character of planetary research during the 1980's will remain largely empirical, although a number of factors should lead to a rapid growth of theoretical and broadly based interpretive investigations. Prominent among these factors is the existence of an extensive and relatively undigested data base from previous and current space missions that now shows signs of becoming reasonably accessible to the community at large.

Two major purposes underlie the multitude of specific research interests: a strong drive to establish a broad empirical basis for deciphering the cosmogony of the solar system and the continuation of vigorous efforts to attain a balanced and more complete exploration of the current state of the planetary system.

We expect these drives to be strongly evident throughout the full range of research techniques employed in planetary science, and we have identified the following broad set of requirements that we believe need to be met if the current level of research momentum is to be maintained.

A. Planetary Missions

While it was not part of the charge to this Working Group to recommend future planetary deep-space missions, it nevertheless seems essential to state our strong conviction that research momentum in planetary science would rapidly dissipate without an active program of such missions. The case has been clearly stated and a well-considered strategy developed by COMPLEX (see earlier sections for a list of recent reports). This Working Group soundly endorses the positions taken by COMPLEX regarding objectives for planetary exploration, advanced instrumentation development, international cooperation in planetary exploration, the need for adequate launch and interplanetary thrusting capabilities, and the importance of a

strong program of ground-based and Earth-orbital observations
to complement exploration by deep-space probes.

B. Adequate Support to Individual Investigators

By far the largest source of funding for individual investigators in planetary science flows from NASA's Office of Space
Science. NSF direct funding for specific solar-system investigations (excluding the Sun itself) is a modest $0.5 million/
year and supports roughly a dozen investigations. The level of
NASA funding--plus advanced mission planning, data analysis,
and long-term instrument development--has remained roughly
constant at approximately $25 million per year throughout the
past decade regardless of the loss in purchasing power of the
dollar. In some areas, such as lunar science, there have been
drastic and unexpected reductions of support. Data analysis
associated with highly successful and continuing missions
(e.g., Pioneer Venus, Voyager at Jupiter) has been, or is
planned to be, severely curtailed. We detect concern in the
planetary-science community that erosion in the level of
support for individual and supporting research is inevitable
and might possibly accelerate in the future if new starts on
planetary missions become less frequent. Yet there are also
clear indications of growth in the community and in a data base
that is becoming more available to many investigators not
directly connected with missions. New technology is providing
increased capability for ground-based and Earth-orbital telescopes, and observers are encouraged to address new and more
difficult problems.
   If the maximum scientific benefit is to be obtained both
from already acquired data and from these new capabilities,
then it seems essential to us that the funding level to basic
research by individuals must be increased in future years. Our
Working Group has not had either an adequate base of information or the time to assess properly the magnitude of the needs
in this area, but it does seem that a significant problem
exists that requires careful examination and action by both the
Office of Space Science at NASA and the relevant Divisions
within NSF.

C. The Use of Existing and Planned Astronomical Facilities

Many of the ground-based and essentially all of the Earth-
orbital facilities currently used by planetary astronomers have
been developed with the primary objective of addressing a wide
range of questions in astrophysics. One important exception to
this has been the successful development of several large-
aperture optical telescopes (the 2.2-m and 3-m on Mauna Kea,

the 2.7-m at McDonald, and the 1.5-m Catalina) by the Planetary
Astronomy Program Office at NASA to serve some of the needs of
planetary science. We estimate that the fraction of available
time that is assigned on non-NASA astronomical telescopes for
solar-system studies is currently between 5 and 10 percent.
The NASA telescopes are devoted approximately 50 percent of the
time to solar-system work. Thus, the numerous ground-based
advances and discoveries outlined earlier can be attributed to
a rather limited amount of observing time.

For the future, we see no slackening of interest by planetary astronomers in existing and planned astronomical systems.
In fact, we anticipate the reverse tendency, especially in the
case of ST. Therefore, it seems productive for us to highlight
some of the requirements that planetary scientists would like
to see satisfied in general-purpose astronomical facilities.

We foresee substantial benefits flowing from advances and
proposed developments in radio astronomy. The high spatial
resolution afforded at a range of microwave frequencies by the
VLA should have important applications to studies of the
structure and composition of the deep atmospheres of Venus and
the outer planets and possibly to their dynamics. At long
wavelengths (greater than 10 cm) the VLA may possibly expose
thermal anomalies, e.g., active volcanism, on the surface of
Venus. This would be an important supplement to morphological
data obtained by radar.

At higher frequencies improvements in receiver sensitivity
and in antenna quality and size present opportunities to improve
our knowledge of asteroid surfaces and the composition of
cometary atmospheres. The proposed 25-Meter Millimeter-Wave
Telescope could lead to a major expansion of our knowledge of
the variety of parent molecules and whatever other molecular
fragments are vaporized from cometary nuclei as they approach
the Sun.

In radar astronomy, the present emphasis--which has been on
topographic and morphological mapping of the surfaces of Venus,
Mars, and the Moon--is expected to shift to investigations of
asteroids and, whenever possible, to cometary nuclei. It is
estimated that roughly 60 known asteroids will be within the
range of current instruments during the next few years, and a
substantial increase in knowledge of their surface characteristics and rotational properties can be expected from an
aggressive program.

In the infrared the potential of ground-based and Earth-
orbital observations seems enormous. Accurate radiometry and
spectrally comprehensive moderate- to high-resolution spectrophotometry in the 1-5-μm region is needed on all the outer-
planet satellites, Pluto, cometary nuclei, and asteroids in
order better to characterize their global properties and
surface compositions. High-resolution imaging spectroscopy is
needed to understand the morphology and stability of outer-

planet stratospheres in the emissions seen at 7.8 and near 12
μm. Suspected changes with time need to be studied.
Similarly high spectral resolution "imaging" studies of individual absorption lines of the many molecules seen in the
5-μm window are needed to reveal the center-to-limb behavior
of the absorptions and any possible lateral inhomogeneities.
Such information is also needed to resolve questions of
atmospheric structure, composition, and mixing processes. At
longer wavelengths (10-100 μm), the thermal spectra of the
outer planets beyond Jupiter are still poorly determined, and
much could be learned about the structure and primary composition of their upper tropospheric regions if moderate-
resolution calibrated spectra were available. The very high
spectral resolution attained by heterodyne spectroscopy near
10 μm has already proven useful in following motions in the
stratosphere of Venus, although more comprehensive investigations are required. Also, the "flare" phenomenon at 5 μm
that has recently been discovered on Io, and that is presumably
related to volcanic activity on the surface, urgently needs
further detailed investigation.

Most of these needs should be satisfied by already proposed
or currently operating programs; for example, the development
and increasing availability of array detectors for the infrared
is clearly important, as is the combination of low background
and good image quality that SIRTF will provide. There seems to
be no question that planetary science would greatly benefit
from the presence of a high-resolution infrared spectrometer on
SIRTF and also ultimately on ST. In addition, IRAS should
provide a wealth of basic radiometric data on many asteroids.

In the optical and ultraviolet spectral regions, it is
already clear that there will be strong demand for use of ST
for solar-system investigations, since it provides roughly a
tenfold improvement over available imaging quality in the
optical region, together with a dual increase of a tenfold
improvement in spatial resolution and a hundredfold increase in
spectral resolution in the near ultraviolet. ST should have an
enormous impact on essentially all problems in planetary
astronomy, particularly in the study of planetary atmospheres
(dynamics, rotation, upper-atmospheric aeronomy, stratospheric
composition and structure) and in its application to
magnetospheric interactions and circumplanetary nebulae at
Jupiter, Saturn, and possibly beyond. Its probable impact on
the study of planetary surfaces is less certain, since little
is known at present about the appearance of surfaces in the
2000-3000-Å spectral region. However, if strong contrasts
exist in this spectral range as a signature of surface chemistry, then images at the spatial resolutions that can be
attained on Mercury, Mars, the Moon, and the Galilean satellites will be of considerable interest. Finally, the great
sensitivity of ST above the sky background should make it an

important tool for studying the chemical properties of the surfaces of large numbers of faint cometary nuclei and asteroids.

We can also foresee the use of ST in conjunction with planetary missions. For example, the full spectroscopic and imaging capability of ST will be of great importance to the Galileo mission and to a future cometary rendezvous mission. While the Galileo orbiter will be capable of making measurements of specific Jovian features at 10 times the spatial resolution of ST and over a much wider range of viewing angles, it will only be able to do this periodically and over limited periods of time. ST, on the other hand, will be able to provide (at least in principle) an essentially continuous record of the development of related large-scale motions in Jupiter's atmosphere throughout the entire Galileo mission. In the case of a comet mission the rendezvous craft would spend most of its time embedded within the cometary atmosphere close to the nucleus making in situ measurements; as a result, the instruments on this spacecraft will experience difficulty in following what is happening in the comet's atmosphere on the largest physical scales. ST, on the other hand, will be an ideal system for providing complementary information regarding these large-scale phenomena.

This synergism between observations from Earth-orbital telescopes and planetary spacecraft has not been clearly recognized in the past, but the potential capability of ST now makes it essential to correct this. We believe that there is some urgency for those organizations that will be responsible for operating and scheduling the Earth-orbital telescopes and those managing space missions to work out management systems that will ensure that adequate complementary observational coverage at the most important times is available.

D. Laboratory and Theoretical Studies

There are numerous urgent laboratory studies that are needed in planetary science if we are to harvest the full value of the remote-sensing capabilities available.

Spectroscopic information under relevant physical conditions is of particular importance. The list of molecules that have been discovered in planetary atmospheres and in cometary atmospheres is now quite long, including H_2, SO_2, CO_2, HCl, HF, PH_3, NH_3, CH_4, C_2H_6, C_2H_2, H_2O, GeH_4, SiH_4, H_2S, CO, HD, NO, O_2, CH_3D, CH, CN, HCN, C_2, C_3, and various isotopic forms. There is a similarly long list of atomic and ionic lines. The transitions that have been used to detect these molecules and atoms are of all types and range from the extreme ultraviolet to microwave frequencies. The remote-sensing approach uses the observed strengths, positions, and (occasionally) the widths of spectral lines to infer physical and chemical properties. Unfortunately, the molecular and atomic parameters that allow these inferences are usually poorly known or else are known only under physical conditions that are not directly applicable; as a consequence, substantial extrapolations must often be made to arrive at a result.

In many cases it is the error associated with a laboratory determination of a line parameter, rather than the uncertainty in a telescopic measurement, that limits the accuracy of a result. There is thus a great need for accurate information on line intensities, pressure shifts, broadening coefficients, and line shapes for all the molecules listed above, at pressures and temperatures and in mixtures that are representative of planetary conditions. This is, of course, not a new requirement, and a number of small laboratory programs have been funded for some time. However, while substantial progress has been made with transitions in a few molecules such as H_2, CO_2, NH_3, and CH_4, this has only been done with precision at normal pressures and temperatures. A few preliminary studies have been made on CH_4 and H_2 absorptions at low temperatures, but progress has been extremely slow.

We have received two suggestions for correcting this situation. One is a proposal to set up a national center for spectroscopy, while the other is to develop the capabilities of one or two of the existing research groups into well-equipped regional centers. At the heart of both of these suggestions is the establishment of a well-instrumented spectroscopic facility to which qualified research scientists could go and acquire (either by themselves or with the help of a resident staff) spectroscopic data of particular interest to them. We have found that both of these ideas are controversial. Nevertheless, they do reach the heart of the problem, and we recommend that they be considered in more detail. One fact is clear: a substantial investment in sophisticated equipment--such as very-high-resolution spectrometers; long-path-length, temperature-controlled, vertical-absorption cells; and high-quality light sources--will be required.

A second area in which continued laboratory effort needs to be supported is reflectance spectroscopy of assemblages of natural minerals and ices. Such information has already led to the remote detection of water ice in the particles that make up Saturn's rings, the presence or absence of water ice on the surfaces of the Galilean satellites, and the presence of SO_2 frost on Io. Similarly, reflectance spectroscopy in the 0.3-2.5-μm spectral region has played an essential role in the classification of asteroid types and the exploration of their possible mineralogical types. Much remains to be done in the future in providing a wider selection of reference spectra and their extension into both the near-infrared (2.5-5 μm) and the near-ultraviolet (0.2-0.3 μm) regions.

In addition to spectroscopic results, physical and chemical data are badly needed. Theoretical progress in understanding the structure of planetary interiors and behavior of volatiles on surfaces or in atmospheres (e.g., cloud physics) requires the acquisition of chemical and physical data on equations of state, phase changes, solubility, vapor pressure, and condensation and nucleation properties of a range of substances and mixtures under relevant physical conditions. For many common substances, of course, much of this information is already available; in many cases, however, existing data must be extrapolated to conditions of interest. For interiors of the planets and satellites, information on mixtures of H_2-He, CH_4-NH_3-H_2O, and refractory materials are required. We understand that such information can be obtained with the aid of high-pressure diamond-cell technology, and we encourage support to investigations that will produce information relevant to planetary science. For understanding observed phenomena on planetary surfaces and in atmospheric clouds and hazes, information on vapor pressure, condensation, and nucleation for S, SO_2, NH_3, H_2SO_4, CH_4, and NH_4SH are urgently needed.

Finally, advanced computing systems hold the key to progress in a number of areas. Efforts to understand primitive conditions in the solar nebula, motions in planetary atmospheres, the details of the complex magnetosphere interactions seen in the Jupiter-Io system, planetary magnetism, and atmospheric evolution will necessitate increasingly complex computer simulations of these phenomena. When performed by qualified scientists, these simulations may provide the only reasonable demonstration of the adequacy of a theoretical explanation of a complex phenomenon and are to be strongly encouraged. In addition, they may also yield otherwise unobtainable physical insights into the mechanisms by which the phenomena occur. We consider that it is important that adequate access to the most advanced and fastest available computing machines be made available to qualified planetary scientists.

E.   Extrasolar Planetary Detection

One of the two major goals that we have identified in planetary science is the establishment of a wide empirical base for deciphering the cosmogony of the solar system. At the center of this research effort will be laboratory, ground-based, and Earth-orbital telescopic and deep-space investigations of primitive bodies in the solar system.

However, this by itself will probably not be enough to provide a complete picture. Contributions from observations of related phenomena in the Galaxy will be necessary to expose the processes that can actually occur during protostellar collapse and the fragmentation of interstellar clouds. Equally important will be the search for and discovery of extrasolar planetary systems that will ultimately reveal the statistics of planetary-system formation.

At present it is generally assumed that the formation of planetary systems commonly accompanies star formation, but in fact no proof exists. There is a large gap between the smallest objects that have actually been detected in binary systems (approximately 0.1 solar mass) and the mass of the largest known planet (about 0.001 solar mass). Knowledge of the statistics of occurrence in multiple and binary systems of objects in this mass range and their relationship to the physical characteristics of the dominant star would be a major impetus to clarifying the way in which planetary systems actually form. The actual statistics of planetary formation will relate directly to the physical status and the relative peculiarity of our own solar system and improve the estimation of the probability of the existence of other life in the Galaxy.

Attempts have been made in the past to detect planetary systems by astrometric means, but these efforts have apparently not attained the required accuracy. An astrometric precision of 10^{-4}-10^{-5} arcsec is required, and this must be available for many stars and maintained over time scales of approximately 10 years. Recent studies in NASA-sponsored workshops have indicated that such precision may now be technically possible with dedicated astrometric facilities. Other methods, which appear to be less certain of success but clearly not impossible, include the detection of small periodicities in radial velocities, spacecraft interferometry, and direct detection by means of specially apodized large-aperture telescopes.

As outlined above, the detection and accumulation of statistics on other planetary systems would be an important development; given the likelihood of reaching the required accuracy by means of astrometric techniques, we support the start of programs in this field. The development that we envisage should include within it at least two independent research programs in order to ensure adequate cross-checking of positive results.

F. Dedicated Orbital Telescope for Planetary Studies

The power of remote sensing from Earth-orbiting telescopes is now receiving wide recognition in planetary science as a result of the successful application of OAO-A2, Copernicus, and IUE observations to a selection of planetary and cometary problems. There is also great anticipation for what will be learned with ST and SIRTF in the future.

However, in spite of the capability of the above facilities to attack many solar-system problems, there are also many drawbacks and mismatches. Many solar-system observations require telescopes to be pointed in the close vicinity of the Sun or of the Earth or require special orientations of a spectrograph slit or complex attitude maneuvers. It is the experience of those responsible for operating existing Earth-orbital telescopes that solar-system observations are among the most difficult to accomplish from an operational point of view. There are solar-system observations that need to be made in spectral regions (e.g., below 912 Å) or with combinations of spectral and spatial resolutions that are not necessarily appropriate for astrophysical problems. In addition, many solar-system observations have either special timing requirements or need extended periods of observations, and the limited availability of astrophysical instruments for such studies is clearly a problem. This could be a particularly exasperating problem when observations are associated with a planetary mission in progress.

Our assessment of these considerations is that, taken together, they have considerable substance, and it is our opinion that special consideration should be given to the development of a dedicated Earth-orbital or Shuttle-based telescope for solar-system studies. However, we also believe that it is essential that the priority of this concept be developed in the context of the overall priorities for an integrated program of solar-system exploration.

G. Occultation Network

Observations of the occultation of stars by planets and asteroids have led to a remarkable number of discoveries in the last decade. Among these are the discovery of Uranus's rings and the presence of high temperatures and turbulence in planetary stratospheres. Occultations also provide spatial information on the phenomena they probe, and returns from two of the most successful occultations--the 1976 occultation of Epsilon Gem by Mars and the 1977 occultation of SAO158687 by Uranus--were greatly enriched because reasonable data were obtained from several stations. In fact, we would have a much better picture of the global pattern of Martian atmospheric tides if the geographical distribution of observing stations for the Epsilon Gem occultation had been broader. Similarly, the dimensions of asteroids and the figures of planets may be refined through occultation techniques.

In order to be prepared to take full advantage of future occultation events, we declare our support for the concept of a loose, worldwide network for promoting occultation observations. Ingredients of such a network would include: (1) some method of centralized coordination of observing plans; (2) the continued development of a good list of predictions for occultation events and their disbursement; (3) the establishment of

adequate instrumentation at a broad geographical distribution
of observatories; and (4) the identification of observers at
each observatory who would be interested in the results.
Finally, interested amateur astronomers would be utilized in
the plans for this worldwide network.

VI. ACKNOWLEDGMENTS

The Working Group would like to acknowledge the interest of the
following individuals who communicated their ideas and
frustrations on the current state and future directions of
planetary science: T. Thompson, G. Downs, R.S. Saunders, J.
Burke, R. Smoluchowski, U. Fink, R. Reasenberg, D. Clayton, J.
Margolis, W. Sinton, J. Apt, D. Black, J. O'Keefe, M. Davies,
H.W. Moos, R. Beer, D. McCleese, and H. Masursky.

3

Galactic Astronomy

I. INTRODUCTION

Galactic studies play a special role in astronomy, affecting
fields as diverse as planetary science and cosmology. The
Galaxy is at once the environment of planetary systems and the
most accessible example of the building blocks of the Universe.
Galactic astronomy in the 1980's will be characterized particularly by the synthesis of many techniques of investigation to
improve our understanding of the evolution of the Galaxy and
its constituents.

Much of the research during the next decade will build upon
the major discoveries of the 1970's. Essential progress in
stellar astronomy has been achieved in our understanding of the
late stages of stellar evolution. Detailed theoretical studies
of the characteristics of collapsed objects and observational
efforts to prove the existence of black holes have been particularly rewarding. Studies of interstellar matter were distin-
guished by the discovery and interpretation of the hot component
of the interstellar medium and by the development of the field
of interstellar chemistry.

The first observations of Galactic regions of star formation, together with detailed observations of coronal activity
and mass loss in stars, have provided compelling evidence that
matter is exchanged between stars and the interstellar medium.
Isotopic studies of chemical abundances in meteorites have been
combined with theoretical work on nucleosynthesis in advanced
stages of stellar evolution to provide evidence that the chemical composition and formation of our solar system may have been
related to the explosion of an early supernova. Prospects for

138

understanding the chemical evolution of the Galaxy have been greatly enhanced by the recent development of observational and theoretical tools for studying stellar abundances.

These recent advances are characteristic of the main activities of Galactic astronomy. They typically involve, at one level, the study of particular classes of objects, each for its intrinsic interest. At another level, these activities are related by our need to understand the complex interrelations among the constituents of the Galactic system.

Many of the processes occurring in the Galaxy combine to form cycles. The relentless process of stellar evolution transforms the primordial gas from which the Galaxy formed into an ever-richer mixture of complex atomic nuclei, molecules, and dust particles. Star formation, nucleosynthesis in stellar interiors, and the injection of the processed material into the interstellar medium form a cycle basic to the chemical evolution of the Galaxy. This fundamental cycle has produced the complex building blocks that form the basis for all living organisms. Dust grains produce conditions favorable for star formation by shielding portions of the interstellar medium from the heating effects of starlight. The grains are incorporated into new stars, and the cycle is completed when new grains formed in the cool atmospheres of evolved giant stars are ejected into interstellar space. Within the interstellar medium there may also be cycles in which grains are created and destroyed. There may be chemical cycles depending on the effects of starlight, cosmic rays, and surface phenomena on grains. Some cycles may involve the creation and destruction of complex atomic and molecular systems. Interstellar gas circulating around the Galaxy may pass through spiral arms periodically causing density waves that may trigger star formation.

The patterns and cycles of Galactic evolution are dimly perceived and only vaguely understood. Many questions must be answered before these processes can be elucidated. What fraction of the Galactic mass is in unseen objects--black dwarfs, planets, collapsed stars? What processes distinguish the Galactic disk from the halo? What conditions are required for stars to form in an interstellar cloud? How is the energy of supernova explosions distributed to cosmic rays and to the motions of interstellar clouds, possibly leading to mass loss from our Galaxy? How are heavy elements formed by nucleosynthesis in late stages of stellar evolution and in catastrophic events such as supernova explosions? What determines the balance of matter in dense clouds, diffuse clouds, and the hot component of the interstellar gas? These are but a few of the unifying questions of Galactic astronomy that must be addressed in the 1980's.

This chapter discusses the kinds of research efforts that are expected to contribute in the coming years to an improved understanding of these important issues in Galactic astronomy.

The following section contains the summary and conclusions of the Working Group. Section III contains a summary of some of the principal scientific goals for Galactic astronomy in the 1980's. Section IV deals with studies of components and subsystems of the Galaxy that may delineate the fundamental processes of Galactic evolution; these studies will provide a basis for efforts to synthesize a comprehensive picture of the origin and evolution of the Galaxy. Section V describes studies of single stars, systems and groups of stars, and the interstellar medium. Observational, laboratory, and theoretical efforts that are essential foundations for continued progress in Galactic astronomy are discussed in the last section.

II. SUMMARY AND CONCLUSIONS

Galactic astronomy has been, and will continue to be, fundamental to understanding virtually all aspects of astronomy. On a large scale, studies of our Galaxy provide the baseline for studies and interpretation of extragalactic objects. A major overlap between Galactic and extragalactic astronomy has been the study of the roles that major subsystems common to our own and other galaxies play in galactic evolution. The next decade holds promise of a new era in which studies of the Galaxy may yield insight into its early history and thereby provide a direct and detailed view into processes that are currently veiled or hidden in extragalactic objects at the edge of the Universe. On a far more modest spatial and temporal scale, studies within our own Galaxy of the frequency of occurrence-- and indeed, the existence--of other planetary systems are necessary for an understanding of the origin of our own planetary system. They would also provide invaluable observational constraints on our understanding of star formation.

The examples cited above are intended to indicate the scope of astronomical problems in which Galactic astronomy is an essential ingredient. In the following sections of this chapter, we identify and discuss a wide range of challenging problems in Galactic astronomy that are judged to be of importance for a further understanding of the Galaxy and astronomy in general. However, pursuant to our charge to identify those problems in Galactic astronomy that appear to be both important and timely for the coming decade, we first call attention to those problems that we believe merit particular consideration during the 1980's.

The Working Group reached a consensus that three of these problems represent truly outstanding opportunities for the coming decade and are worthy of the highest priority. These are, without regard to order of importance, the following:

* The formation and evolution of molecular cloud
complexes and their relationship to star formation and spiral
structure;
* The dynamics and evolution of star-formation regions
and protostars, including the interaction of supernova remnants
with these regions; and
* The formation, evolution, and properties of compact
stellar objects.

In addition, the Working Group identified six problems that we
believe represent excellent opportunities for the 1980's,
although as a class we rank them lower in priority than those
above. These are, again without regard to order of importance,
the following:

* The formation and evolution of the protogalaxy;
* The nature and composition of the Galactic halo and the
Galactic center;
* The chemical composition of interstellar and
circumstellar gas and dust;
* The dynamics and thermodynamics of the interstellar
medium and the interaction of supernova remnants with the
interstellar medium;
* The stellar and dynamical evolution of binary systems;
and
* The advanced nuclear-burning stages of massive stars.

We stress that the nine opportunities given above are not felt
to be necessarily more important than the others mentioned
later in this chapter. Rather, they are the ones that were
judged to be both important and particularly timely for the
1980's.

III. THE LARGE-SCALE STRUCTURE OF THE GALAXY

A. The Galactic Ecosystem

Galactic astronomy involves the continuing study of the
Galaxy's extraordinary variety of constituents and wide range
of environments. These studies have a unifying theme: the
interrelation of these components in a Galactic ecosystem whose
balance and evolution is perhaps as complex and delicate as
that of a tidepool or a tropical forest. Viewing our Galaxy as
the synthesis of its parts provides the link between Galactic
and extragalactic astronomy. What does the spectrum of a
distant, unresolved galaxy imply about its internal stellar
populations? Why do some galaxies spawn young stars while
others do not? How do galaxies evolve? How are their characteristics influenced by their extragalactic environments?

These are questions about the processes in other galactic
ecosystems, and it is unthinkable that we could develop satisfactory answers to them without at the same time understanding
these processes in our own Galaxy.

We have some understanding of a few of the cycles within
our Galaxy: the flow of gas from dense interstellar clouds to
stars and its return to interstellar space; the cycle of interstellar grain creation and destruction; and the chemical cycles
establishing the balance of interstellar molecules. We have
ideas of how to deduce the past history of our Galaxy by measuring the nuclear residues of stellar evolution and cosmic-ray
irradiation and by counting collapsed stars. This understanding, however, is emerging only in dim outline; many details and
vital links in the logical chains are lacking. How much mass
is in objects as yet unobserved such as black dwarfs, planets,
or collapsed stars? What physical processes and populations of
objects distinguish the Galactic disk from the halo? What are
the processes that trigger star formation in an interstellar
cloud? How is the energy of supernova explosions imparted to
the interstellar medium? What determines the balance of matter
in dense clouds, diffuse clouds, and hot interstellar gas? The
answers to these questions are keys to understanding the order
of the Universe.

B. Size and Mass of the Galaxy and the Galactic Halo

Among the fundamental quantities of Galactic research in need
of revision are the distance of the Sun from the Galactic
center (which now appears to be less than the 10 kpc usually
adopted) and the rotation velocity of the solar neighborhood.
Correct values of these quantities are crucial to our knowledge
of the scale and mass of our Galaxy. Our understanding of the
properties of the Local Group of galaxies also depends on these
quantities. Basic questions concerning the kinematics of the
Galaxy are currently under debate. The debate includes renewed
discussion of a possible expansion of the young stellar and
gaseous components of the Galaxy, and the role of noncircular
motions in the Galaxy. These matters all relate to determining
the Galactic rotation curve, which expresses the velocity of
rotation as a function of distance from the center of the
Galaxy.

A remarkable development is the recent discovery that the
rotation curve of most spiral galaxies remains flat to the
limits of detection; in fact, there have been no normal galactic rotation curves yet reported. This result implies that the
total size and mass of spiral galaxies are much larger than
previously thought. Moreover, because most of the material in
the outer region is unseen, the physical properties of this
material are largely unknown. Although the possibility that

the unobserved mass is gas appears to be ruled out, there is still no evidence for large numbers of faint M dwarfs in the halo. The properties of the unseen material may differ from those of the known stellar and interstellar populations. Until recently, the kinematics of the outer regions of our Galaxy were unknown; now, spectra of stars and of regions of ionized gas, together with radio observation of molecular regions, are producing detailed results.

Considerations of the stability of galaxies and interpretations of the rotation curves may require that the visible components of spiral galaxies be embedded in massive halos. Because such halos are difficult to detect and observe, dynamics must play an important diagnostic role in their study. From this point of view, the observed components of a galaxy constitute test bodies that may be used to probe the total gravitational field of the system.

C. Dynamics and Spiral Structure

During the past decade, major efforts were made to develop a dynamical interpretation of spiral structure in galaxies. By 1970, the theory of spiral density waves could provide a tentative physical picture of spiral structure that lent itself very well to the interpretation of a variety of optical and radio observations. Many such applications of the theory have been made. But it also became apparent during the 1970's that spiral structure could well be a more complicated phenomenon than was assumed in the early versions of the density-wave theory, and it became particularly clear that the excitation of such waves (or other disturbances) in a galaxy was a serious theoretical problem. The problem of the origin of spiral structure has led to systematic theoretical investigations of unstable modes of oscillation in galactic disks, the response of galactic disks to external perturbations (e.g., by companions or intruders), the gravitational forcing of spiral structure in the interstellar gas by bars or large-scale oval distortions in a galaxy, and the role of resonance phenomena in the dynamics of spiral waves. Future theoretical work on spiral structure should take into account the self-consistent behavior of stars (including resonance effects and possibly other nonlinearities); nonlinear and dissipative processes in the interstellar gas; and the exchange of matter, energy, and angular momentum between the stars and gas.

D. The Galactic Nucleus

Galactic nuclei in general--and, because of its proximity, especially the nucleus of our own Galaxy--provide exciting opportunities for studying stellar dynamics and astrophysics. The nuclei of normal galaxies, including our own, show evidence of violent activity, high-velocity dispersions, geometrical asymmetries, mass ejection, and a central mass of high density. Such effects can be studied in great detail in the nuclear region of our own Galaxy.

The central region is currently the subject of observations at many wavelengths. Continuum and spectral-line radio data provide evidence that suggests either the violent ejection of matter or the presence of a barlike structure. Infrared and radio data currently suggest a very high central mass concentration a few parsecs in diameter that may contain a black hole. Gamma-ray observations are providing constraints on the cosmic-ray density that will add information on the nature of the core region of the Galactic center.

The nuclear region of our Milky Way Galaxy has an infrared luminosity more than 10^9 times the solar luminosity concentrated in a region a few parsecs across. Although the nature of the energy source is uncertain, there is evidence that both thermal and nonthermal processes play a role. In addition, preliminary infrared maps reveal complex spatial structure in the region surrounding the Galactic nucleus. One suggestion is that the nucleus contains numerous planetary nebulae. In any case, the powering source must consume about 10^{-4} solar masses/year to provide the observed luminosity; if the source has existed since the period of Galactic formation, 10^5-10^6 solar masses have now been consumed. It is difficult to provide this large luminosity through ordinary stellar nuclear-burning processes; a nonstellar process, such as the consumption of matter by a massive black hole, could be involved. The high sensitivity and moderate spatial resolving power of a cooled 1-m-class infrared telescope in space promises to reveal a great deal of useful information about the nature of the energy source in the Galactic center.

Galactic nuclei represent a special regime in galactic dynamics, involving relatively new problems concerning the equilibrium, stability, and evolution of stellar systems. Observational studies of our Galactic center and of the central regions of other galaxies require the theoretical consideration of dense concentrations of mass at the centers of galaxies. The circumstances may be such that stellar encounters, and even stellar collisions and mergers, play an important role. In extreme cases, the effects of General Relativity are relevant. The gravitational and hydrodynamical coupling of the stars and the interstellar gas may be significant. The gas distribution in the inner few kpc of the Galaxy is evidently tilted with respect to the Galactic equator but nevertheless shows substantial spatial and kinematic coherence and symmetry.

E.   Globular Clusters

Globular clusters are of essential value to the study of Galactic astronomy because of the relative simplicity of their structure and dynamics, the extreme character of their stellar population, and the information they provide about the dynamical and chemical evolution of the Galaxy.

Refinements during the 1970's in photometric studies, star counts, measurements of proper motions, and measurement of radial velocities have led to substantial improvements in our knowledge of the structure and kinematics of globular clusters. Significant tests of equilibrium models of clusters using these data should be possible during the 1980's. Studies of clusters have shifted in emphasis from the problem of their equilibrium to the problem of their evolution. Recent studies have led to an evolutionary picture that suggests the formation of a core-halo structure whose primary features are a contraction of the core and an expansion of the halo; the release of energy in the binding of binary stars must play a substantial role in these processes. Although the initial phases of the core-halo evolution of clusters seem well established, understanding the ultimate fates of the globular clusters will require much further study. The loss of stars through evaporation and tidal disruption must be studied to assess the importance of these processes. The process of core collapse must be examined to determine whether massive objects can form. It remains to be understood whether mass loss from stars modifies the processes of relaxation, equipartition of energy, and the stratification of stars of different masses in a cluster.

Substantial advances have occurred in studies of chemical abundances, and it now appears that there are significant variations of age and chemical composition among the globular clusters. These results will play an important role in future efforts to reconstruct the evolution of the Galaxy. They already suffice to show that the initial, rapid collapse of the halo that has been envisaged in the past oversimplified what must have occurred during the formation of the Galaxy. Observations of the fainter constituents of globular clusters--especially close binaries, white dwarfs, and other compact objects--will contribute substantially to the study of dynamics and stellar evolution in these systems. Comparisons of globular clusters with open clusters and with clusters in the Magellanic Clouds should provide important insights into the formation and early evolution of clusters.

The discovery during the 1970's of x-ray sources in globular clusters is especially exciting because of the possibility that dynamical evolution could lead to the formation of massive objects (such as black holes) in the cores of clusters. Alternatively, these x-ray sources could be representative of

late stages of stellar evolution. Whatever their nature, the
x-ray sources raise important new questions to guide the study
of globular clusters during the 1980's.

F. The Galactic Environment: Warps, High-Velocity Clouds,
and the Magellanic Stream

Many aspects of the outer regions of our Galaxy should receive
intensive study during the coming decade. Earlier radio work
has revealed an extensive distribution of high-velocity clouds
of gas covering large portions of the sky. Subsequent work on
our own and other galaxies has shown that their surrounding
environments are not always empty intergalactic space, but
rather that many galaxies are accompanied by remote globular
clusters, dwarf galaxies, and isolated clouds of gas as well as
by extensive streamers of gas with length scales greater than
the dimensions of the host galaxies. Although together with
the Galaxy itself these diverse phenomena form a single interacting ensemble, the dynamical consequences of the interactions
are unknown--as are, indeed, the origin, total mass, and
detailed constitution of the environmental material. It is not
yet known, for example, whether the high-velocity clouds and
the Magellanic Stream represent gas whose origin is tidal or
primordial. Whether the various streams of high-velocity gas
are related and the extent to which this gas interacts with the
Galaxy at large are questions that future studies will have to
answer.
    Because the total content and nature of this environment is
unknown, the dynamical influence of the environment on the
Galaxy at large is difficult to study quantitatively, although
this influence is certainly important and pervasive. There is
some evidence to suggest that some of the streams impinge on
the Galactic disk. The large-scale gas flows that would result
would play a key role in the chemical evolution of the Galaxy.
Such inward flows, or eventual gas flows of a different origin
outward from the Galactic nucleus, could determine how elements
made by stars are transported through the Galaxy and how they
are diluted by less enriched gas. Enough is known about radial
flows in the disk, flows to and from the halo region, loss of
gas in winds, and gain by accretion to be sure that these
processes are vitally important in the evolution of the Galaxy.
Stellar and interstellar abundances themselves suggest that
infall of metal-poor gas occurs at a rate comparable with the
star-formation rate. Only direct observational and theoretical
studies of large-scale gas dynamics can provide the detail
needed for a full understanding of these phenomena. Progress
in the field of chemical evolution also requires such studies,

and they are fundamental to understanding the dynamical evolution of the Galaxy.

Challenging dynamical problems must furthermore be overcome before the warped nature of the outer parts of the Galactic disk are understood. Many external galaxies show the same nonplanar outer structure. Although the dynamics of galactic encounters can explain some warps, other warps are observed in isolated galaxies with no obvious companion. Problems of maintaining the warp against dispersive effects are particularly puzzling.

IV. PROCESSES WITHIN THE GALAXY

A. Mechanisms for Star Formation

Star formation is a process on which almost all aspects of galactic evolution hinge. Theoretical and observational research on all aspects of star formation will continue to be vital, not only to understanding the process itself but also to understanding the structure, chemical evolution, and stellar populations of our own and other galaxies.

Many mechanisms for star formation in molecular clouds have been suggested. The suggestions include supernova-induced star formation as well as star formation associated with ionization fronts and cloud collisions. Theoretical investigation of these processes and comparison of theory with observations is an important area in which substantial progress has been lacking. Our Galaxy is the only galaxy for which sufficient spatial and spectral resolution and sensitivity can be obtained for studying any of these processes in detail; this is particularly true of those processes that involve molecular-line radiation. Results obtained in our Galaxy improve our understanding of other spiral galaxies.

A subject of particular importance in recent and future research is the relationship between the spiral structure in a galaxy and the processes of star formation in the galactic disk. When interstellar gas participates in a density wave or suffers forcing by a bar or oval distortion in a galaxy, shocks may occur that trigger star formation along the spiral arms. An alternative mechanism is that supernova explosions occur stochastically across the galactic disk and trigger star formation. The systems of young stars and H II regions so formed are then sheared into the observed spiral patterns by differential galactic rotation. The wide range of mechanisms relevant to star formation comprise a rich field for further development of theoretical ideas and for a vigorous interaction of theoretical and observational work.

B. Factors Affecting Star Formation

Galactic evolution requires for its understanding more
observations and deeper theoretical insights about the physical
processes that influence the mass function of star formation.
Preliminary evidence suggests that the star clusters in the
Magellanic Clouds, which have had a different evolutionary
history from those in our own Galaxy, have mass spectra that
differ from those observed in Galactic clusters. Furthermore,
the luminosity function of cluster stars differs systematically
from that of field stars. The initial mass function should be
studied as a function of age and chemical composition. We must
also gain insight into the physical processes occurring in
collapsing clouds. Substantial amounts of time on large,
wide-field telescopes should be devoted to these studies during
the coming decade.

C. The Formation and Evolution of the Galaxy

Much effort is currently being devoted to understanding the
origin and evolution of galaxies in general. However, the most
detailed tests of theories of galactic evolution will be provided by studies of the stellar content of our own Galaxy.
This is so because any reconstruction of Galactic history, or
even a direct test of a model, requires detailed information on
the ages and chemical compositions of large numbers of individual stars belonging to different stellar populations. Such
information will, for the foreseeable future, be available only
for our own Galaxy and its closest neighbors.
    Accurate information on stellar ages is particularly impor-
tant for determining how rapidly our Galaxy formed and how many
stages or events were involved in its formation. Recent theories have suggested multiple-stage formation processes or even
formation from several once-independent units. Information on
stellar ages and compositions is only now beginning to attain
the level of accuracy necessary to constrain such theories.
Further theoretical and observational studies to provide accurate data on the history of our Galaxy are of the highest
importance for Galactic research.
    Studies of the chemical evolution of the Galaxy require
data on stellar abundances as a function of time and position
in the Galaxy. These abundances must be measured in unevolved
stars whose surface composition reflects that of the interstellar medium from which they formed; measurements of abundances
in the present-day interstellar medium are also required. Currently we know only the broad trends of metal abundance, which
are observed to increase from the halo to disk stars and
decrease with distance from the Galactic center. Furthermore,
there are a few enticing hints that the relative abundances of

some elements vary systematically with age and/or location. A much larger body of data, giving relative abundances in stars of different ages and positions in the Galaxy, is needed for a sound understanding of these phenomena. Among the most pressing needs is the need for knowledge of abundances of the most common elements (He, O, C, N), about which extremely little is known compared with the case of iron. With more information on the abundances of different elements, one could perhaps understand how nucleosynthesis depends on the masses and initial compositions of stars and how efficiently stellar ejecta are mixed into the interstellar medium. Thus, studies of stellar interiors, star formation, and interstellar gas dynamics would benefit directly from detailed work on chemical abundances.

The small-scale dynamics of our Galaxy are poorly understood. Various types of noncircular motions are observed on scales of different size, but the relations between them are unclear, as are their origins in most cases. Some small-scale motions are probably caused by stellar energy sources such as supernovae, but others have too large a scale size and too much energy to be plausibly ascribed to a single supernova. An example is the mysterious local disturbance in the gas layer associated with the Gould Belt. Some noncircular motions may be due to density-wave effects, but no density wave has yet been unambiguously identified in our Galaxy. Noncircular motions on the scale of 1 kpc or less are particularly important for star formation, since motions on this scale may be involved in the formation of molecular clouds.

There is strong interplay between studies of chemical evolution and star-formation rates at different epochs in the history of the Galaxy. The age distribution of stars in various parts of the Galaxy gives the only direct information about the past history of star formation; without this information, the chemical evolution of the Galaxy cannot be understood. In the last few years, several studies have suggested that this history may not conform to earlier thoughts: globular clusters apparently show an age spread of several billion years; almost all disk stars near the Sun are much younger than any halo stars; and the outermost periphery of the disk contains stars that are both young and metal-poor. It thus appears that even the halo of the Galaxy may have formed through a relatively slow collapse and that the outer disk formed later, with the periphery remaining even now in a very early stage of evolution. Such a picture can be understood qualitatively in terms of recent dynamical models of galaxy formation, but confirmation and clarification require a great deal more data. For example, detailed photometry of stars is needed to infer adequate age estimates from color-magnitude diagrams. The effort of extending current statistical surveys well beyond the solar vicinity would provide unique and essential information on the history of star formation, and indeed on the formation of the Galaxy itself.

D. The Galactic Magnetic Field
and the Confinement of Cosmic Rays

Magnetic fields evidently play an important role in the overall dynamics of the interstellar medium and in the collapse of gas clouds to form stars. The magnetic field of the Galaxy is widely believed to be generated continuously by a dynamo mechanism involving the differential rotation of the interstellar gas, the development of hydromagnetic instabilities, and the escape of flux from the disk. Although some progress has been made in understanding the operation of these processes in uniform models of the interstellar medium, work on more realistic, "clumpy" models of the interstellar medium is only now beginning. Magnetic-field generation in inhomogeneous models is likely to become a focus of attention during the next few years.

An important astrophysical question is the volume filled by cosmic rays and the nature of the mechanisms (presumably involving the Galactic magnetic field) that confine them to this volume. Because cosmic-ray particles propagate along field lines, determination of the confinement volume and confinement time will also yield information on the configuration of Galactic magnetic fields. Key observations include high-precision measurements of isotopic abundances, determination of the ratio of secondary-to-primary cosmic rays at energies of about 100 GeV/baryon, and measurement of the positron spectrum up to 100 GeV. Gamma-ray astronomy will play a major role in answering these questions.

A large cosmic-ray flux in the unexplored region at energies below about 0.02 GeV could play an important role in heating and ionizing the interstellar medium and could also contribute a major portion of the total interstellar pressure. The effects on star formation and the evolution of the Galaxy could be profound. This question can only be settled definitely by direct measurements of cosmic rays outside the solar-wind cavity, because the solar wind excludes these particles from the inner solar system. Before such data are available, some information on the flux of low-energy cosmic rays in the more central parts of the Galaxy can be obtained from a study of the medium-energy gamma-ray diffuse emission and from searches for nuclear gamma-ray lines from, for example, carbon and oxygen.

E. The Hot Component of the Interstellar Medium

Observations from spacecraft during the 1970's of diffuse soft x rays and ultraviolet absorption lines due to interstellar O VI demonstrate that a substantial fraction of the local interstellar medium is filled with hot (about 10^6 K) low-density gas, whose scale height is so great that it merges with the Galactic corona and wind. The presence of this hot gas has important implications for the dynamics of spiral arms, the nature and distribution of interstellar clouds, the propagation and acceleration of cosmic rays, and the interaction of our Galaxy with its intergalactic environment. Its existence is probably related to high-velocity interstellar shocks caused by mass loss from stars in supernova explosions and stellar winds.

What fraction of interstellar space is occupied by the hot gas? How is the gas related to the high-velocity shocks and to the embedded gas clouds? Our ability to answer these questions empirically is limited by our ability to obtain high-resolution UV absorption spectra of faint stars. Space Telescope (ST) will greatly extend this ability, but it lacks capability to observe the O VI lines at 1035 Å, the key indicators of the hot gas at distances greater than a few hundred parsecs, where interstellar opacity precludes direct observations of soft x-ray emission. Recent observations of spectral features arising from oxygen and iron in the soft x-ray background demonstrate the possibility of investigating the physical processes relating the hot gas to the cooler clouds and to the energy sources of the interstellar medium.

V. STELLAR BIRTH, DEATH, AND EVOLUTION

A. Composition of Interstellar Matter

Our picture of the interstellar medium has undergone substantial changes in the past decade. Two major components--the hot component of the interstellar medium and dense molecular clouds--are now seen to be of great importance. Substantial observational and theoretical understanding may be expected in the next decade through quantitative studies of the physics of these two components and their relationship to the major problems of energy balance in the interstellar medium and regions of star formation.

The complex chemical composition of the molecular clouds in the Galaxy will continue to be investigated. With the identification of approximately 50 gas-phase molecules during the 1970's, this area will in the next decade reach maturity, with emphasis on a quantitative understanding of the chemistry. Interstellar chemistry in the 1980's will require substantial theoretical development and, it is to be hoped, attract experienced chemists into the field. Progress on identification of new species will probably best be served by publication of all unidentified lines already found, opening the identification problem to spectroscopists and chemists outside the astronomical community. Exciting advances may include finding the ring molecules, as yet unidentified in even the densest interstellar clouds; a factor of 3-10 increase in millimeter-wave sensitivity will be required for a meaningful search for these.

The chemical indentification of interstellar dust remains a major unsolved problem. The improvement of infrared-detection techniques during the next decade should allow us to make significant new direct observations of the dust in the interstellar medium. A cold IR telescope could gain a factor of 10^3 in sensitivity over existing ground-based IR telescopes, and it will allow us to investigate the extinction laws over a wide range of wavelengths and over enormous distances (10 kpc and greater). A crucial point here is that dwarf-type and subgiant-type stars, whose intrinsic infrared colors are fairly well known, could be identified spectrally and measured photometrically at great distances. Our current understanding of interstellar infrared extinction depends on measurements of luminous supergiants whose spectra are often contaminated by circumstellar features. Measurements of the infrared interstellar polarization can also be made with a cold-telescope facility. These, combined with knowledge of the wavelength dependence of the extinction law, could yield important information about the geometric properties, sizes, and chemical composition of the interstellar grains as well as about the physical conditions of the medium in which they are embedded. For example, these measurements may provide the basis for new assessments of chemical-abundance gradients in our Galaxy and information about the Galactic magnetic-field structure and strength over large distance scales.

B. Molecular Clouds: Sites of Star Formation

The development and maturation of infrared and millimeter-wave astronomy has made possible the first direct observations of the dense component of interstellar matter containing regions of star formation. These areas promise to be an important part of astronomy in the coming decade. Observations of individual centers of star formation with high spectral and spatial resolution ($1-10^3$ AU) make possible exploration of the kinematics and dynamics of the star-formation process. On a larger spatial scale, observations and mapping of molecular clouds are revealing the relationship between dense clouds and young stars. The kinematics and mass of star-forming dense clouds are being studied quantitatively on a galactic scale. In the period from 1955 to 1975, advances in radio astronomy (wavelengths greater than 1 cm) and recombination-line studies led to an increased understanding of H II regions. The emphasis has now shifted to studies of H_2, both directly and indirectly through millimeter-wave observations of CO and other molecules, which are major components of star-forming clouds.

Substantial observational insight into star formation will be gained from studies of two identified types of objects: the infrared-emitting star-formation regions typified by the core of the Orion molecular cloud and molecular cloud complexes containing over 10^5 solar masses.

1. Molecular Cloud Complexes

As a result of observations of CO emission throughout the Galaxy and nearby local OB associations, a picture of inter-stellar clouds in the Galaxy dominated by large massive complexes containing $1-10 \times 10^5$ solar masses is beginning to emerge. These molecular cloud complexes, the most massive objects in the Galaxy, are believed to be sites of current star formation. There are estimated to be up to 5000 such complexes in the Galaxy.

There are many interesting theoretical and observational possibilities raised by the existence of these clouds, includ-ing the relationship of the molecular-cloud complexes to spiral arms and OB associations; the formation, age, and evolution of the clouds; the internal dynamics of the clouds; and the stability of the cloud complexes. The internal velocity dis-persion in molecular-cloud complexes is highly supersonic, and the origin and maintenance of these velocities is an outstand-ing problem.

2. Molecular Cloud Cores, Regions of Star Formation

The discovery of vibrational emission from hydrogen molecules at a temperature greater than 1000 K near the core of the Orion molecular cloud provides evidence that high-energy dynamical phenomena are associated with young stars. Subsequent high-spectral-resolution infrared observations of CO, H_2, and ionized gas leave little doubt that a shock front is moving out from a central source with a velocity of 30 to 50 km/sec and is about 10^{17} cm from that source. Millimeter observations of broad CO emission also indicate an outflow, and the total energy involved has been estimated to be greater than 10^{47} ergs. Thus, an evolutionary process on a time scale of about 1000-3000 years is taking place inside the molecular clouds. The observation of energetic events associated with young stars is an exciting research area in which rapid advances are now possible.

Future observations that take full advantage of new tech-niques may yield observations of protostellar collapse, which, in spite of much effort, has not yet been observed. This is an important area of research and particularly necessary for comparison with theories of star formation, which require more observational guidance.

The development of adequate theories to explain the inter-action between energetic phenomena, such as stellar winds, H II regions, and supernovae with dense clouds, is another promising area of research. The evolution of the dense clouds and the energetics of the interstellar medium in general are strongly influenced by these phenomena. Gamma rays should reveal evidence concerning the interaction of cosmic rays with clouds and may elucidate the role that cosmic rays play in cloud formation.

3. Mechanisms for Star Formation

The internal dynamics of molecular clouds are closely related to star formation. Fragmentation of an initially smooth and quiescent cloud now seems an inappropriate model, and it is probable that turbulence, shock compression, and/or accumulation processes play important roles. Moderate to high spatial resolution millimeter-wave observations combined with infrared and optical studies of young stellar objects will be required to clarify the role of these processes. Multidimensional hydrodynamical calculations capable of representing shock compression, gravitational condensation, and various hydrodynamical instabilities may be required. Such instabilities may turn out to be important for the formation of protostars. Observations will continue to be critical in guiding the parameters that constrain models.

Systematic infrared and optical studies of young objects will be required to correlate their properties with the regions in which they form. For example, there are suggestions that the stellar mass spectrum differs in different regions of star formation, but few systematic studies have been made to identify the factors influencing the mass spectrum. Similarly, more complete data on the frequency of binary stars in various locations or groups of stars will be useful in clarifying the relative importance of fragmentation processes as compared with other mechanisms of star formation.

Both theoretical arguments and observational evidence indicate that substantial quantitites of magnetic flux are lost when interstellar clouds collapse to form stars. The flux may be left behind as a result of ambipolar diffusion, or the magnetic field may be compressed and amplified until it finally disconnects from the interstellar field. In the latter case flux must be destroyed by local processes operating within the collapsing cloud. If the magnetic field of the cloud does disconnect from the interstellar field, the disconnection could occur either in the intercloud medium (leading to the production of cosmic rays there) or in the cloud itself (leading to observable bright spots uncorrelated with variations in the local gas density). Theoretical models of cloud collapse that include magnetic fields have recently been developed. Thus the problem of how collapsing clouds lose magnetic flux in the process of forming protostars may be subject to observational constraints in the next decade.

4. Scale of Observations

A concerted effort to study the spatial and spectral characteristics of H II regions and molecular clouds can be expected
to yield a fairly complete basis for accurate modeling of the
star-formation process within the next decade. These observations are time consuming since they require mapping; substantial
observing time on intermediate-size single antennas, as well as
on arrays, will thus be necessary. Single millimeter-wave
antennas of moderate size with high-sensitivity receivers are
powerful tools for studying overall cloud kinematics, star-
cloud interactions, and thermal balance; they are useful probes
down to distance scales of about $1-2 \times 10^{17}$ cm at wavelengths
of about 1-2 mm.

Analysis of the dynamical motions of the gas and dust in
more highly condensed young objects will require moderate- to
very-high-resolution (10^3-10^5) infrared spectroscopy carried
out with large-aperture telescopes. Radio observations using
large single dishes as well as interferometric techniques will
be required. Infrared surveys from space, such as the Infrared
Astronomy Satellite Explorer (IRAS) program, will be necessary
to locate the more heavily obscured regions of star formation.
A cooled telescope facility may be critical for extending these
observations to extended spatial regions and to distant reaches
of our own and other galaxies.

C. Cosmic Rays

As one of many forms of radiation originating in stellar and
interstellar processes, cosmic rays provide an important
channel of information about these processes. For example,
energetic cosmic rays provide direct evidence of nucleosynthesis and particle acceleration outside the solar system. Thus,
measurements of cosmic rays at the Earth can provide clues to
the aging of stars and to the nature of their deaths, as well
as to the characteristics of violent events near stars and in
the interstellar medium. The Galactic cosmic-ray intensity and
energy spectrum away from the solar vicinity can be deduced
from the study of diffuse Galactic gamma radiation produced in
cosmic ray/interstellar gas collisions. The cosmic rays are
also of great interest in their own right, as a major component
of the Galaxy.

1. Element Synthesis

The theory of stellar nucleosynthesis has been successful in
explaining the relative abundances of nuclei observed in the
solar system. However, elemental abundances in the primary

cosmic-ray spectrum differ significantly from solar-system abundances. These differences could reflect different origins or possible atomic effects in the acceleration process.

Further theoretical work is needed to clarify predictions concerning stellar nucleosynthesis. While adequate nuclear-reaction networks are now largely available, quantitative modeling of element production during the late stages of stellar evolution and during the deaths of stars (supernovae and stellar collapse) is just beginning. With adequate support, substantial progress on these problems may be expected during the next decade.

Measurements of the relative abundances of isotopes in primary cosmic rays provide information on the conditions under which nucleosynthesis takes place, since they are unlikely to be altered during the acceleration process. Supernova explosions are believed to be the site of the rapid neutron-capture process (r-process), which accounts for about half of the solar-system material with atomic number greater than 30. Observations of gamma-ray lines from radioactive nuclei so formed would be a direct ratification that this process occurs in supernovae.

If cosmic rays are strongly enriched in r-process elements, as has been suggested, they will provide a set of r-process abundances possibly characteristic of recent supernovae with conditions different from those that contributed to the solar-system mix.

2. Particle Acceleration

Recent gamma-ray observations lend strong support to the prevailing theory that the bulk of the high-energy cosmic radiation observed near the Earth originates in our Galaxy. However, the process or processes by which synthesized elements are accelerated to cosmic-ray energies is poorly understood at present. Supernova explosions, pulsars, and interstellar shock waves are all considered possible sites for particle acceleration, although significant contributions from stellar flares and nova outbursts cannot be ruled out. This is a complex issue, and no sudden breakthrough is expected during the next few years. However, further theoretical and observation work should clarify the characteristics of these various acceleration processes.

A definitive answer to the question of whether some cosmic rays are accelerated in supernova explosions could come from gamma-ray astronomy. The detection of gamma rays from even one young supernova remnant would demonstrate acceleration of nuclei in supernova explosions, just as detection of synchrotron radio emission has demonstrated acceleration of electrons in supernova explosions. Gamma-ray detectors with improved sensitivity and angular resolution will be required to distinguish such cosmic-ray produced gamma rays from those that may be produced by neutron stars within young remnants. Measurements of elemental energy spectra at high energies can provide important constraints on acceleration models, while measurements of the isotopic composition of heavy elements (such as titanium, cobalt, and nickel) can help to determine the time between the synthesis of these nuclei and their acceleration to relativistic energies. Similarly, measurement of the elemental composition of the actinides can provide information on the time that has elapsed since these elements were synthesized, regardless of when they were accelerated, while measurements of the abundance of isotopes produced by K-capture on elements that are principally fragmentation products can reveal the length of time that cosmic rays spend at low but suprathermal energies before being accelerated. Finally, the study of heavy-element synthesis in solar flares will permit a better evaluation of stellar flares as a site of nucleosynthesis and acceleration.

D. Protostars

Star formation is a very inefficient process; most of the material that reaches the dark-cloud phase of evolution is prevented from evolving into a pre-main-sequence stellar object. This transition phase between dark-cloud and pre-main-sequence star is often referred to as the protostellar phase.

In spite of the obvious importance of the protostellar phase of evolution, the fabric of observational knowledge about this phase is filled with gaping holes. This lack of knowledge is due in part to the fact that the protostellar phase is relatively brief and in part to the fact that, until recently, the wavelength regimes best suited to studying protostars have been only marginally accessible. The coming decade appears to offer the possibility for major advances in our understanding of star formation on both observational and theoretical grounds.

The most likely areas for observational advances involve the early and late stages of protostellar evolution. Because protostars are intrinsically cool objects with an abundance of molecules and because they are relatively small, it is imperative that studies both at high spectral resolution and high spatial resolution be conducted simultaneously. High spectral resolution would be invaluable for establishing the detailed dynamical state of these systems, thereby permitting inferences of the dominant physical processes occurring in protostars and elucidating the interplay of these processes in determining whether evolution can proceed to the stellar state. High spatial resolution is needed not only because of the small size of protostars but also because most of the important processes have characteristic length scales that are smaller than the overall system length scale and are anisotropic. Infrared and submillimeter observations from space offer the greatest promise for understanding the early stages of protostellar evolution.

A somewhat novel observational attack on the late stages of protostellar evolution involves methods that could be employed to search for other planetary systems. Recent studies have examined the possibility of detecting other planetary systems as well as how the results of a search for such systems could place valuable constraints on our understanding of the star-formation process. These constraints are both qualitative as well as quantitative. Current hypotheses regarding star formation imply that planetary-system formation is a natural by-product of the star-formation process. An observational determination of the frequency of occurrence of planetary systems would thus provide a much-needed qualitative test of those hypotheses. If many other planetary systems are dis-covered, their gross properties (e.g., mass, orbital features) could provide a quantitative basis for characterizing the physical state of the late phases of protostellar evolution.

Theoretical and numerical studies of the protostellar state, though ahead of their observational counterparts, are still rudimentary. For example, it is not known at present whether stars form by the free-fall-like, quasi-spherical ac-cretion of material onto a stellar core or by the dissipation of a highly flattened, rotating disk; qualitatively different structures would evolve from these two mechanisms. The loss of angular momentum and magnetic field during cloud collapse are not well understood theoretically, and continued effort in these areas is strongly needed. The complexity of physical and chemical processes involved in protostellar evolution requires that a major emphasis be placed on numerical studies. Support of theoretical and numerical work in this area will be required to strengthen our efforts to understand star formation.

E. Mass Loss, Star-Cloud Interactions

1. Mass Loss in Early-Type Stars

One of the outstanding advances in stellar astronomy during the past decade has been the realization that the early-type (OB) stars lose a substantial fraction of their mass in powerful stellar winds. This result has major consequences for star formation, the interpretation of compact and extended H II re-gions, stellar evolution, supernova theory, the interpretation of supernova remnants, and the evolution of x-ray binary systems.

Observations of stellar winds in the radio, infrared, optical, UV, and x-ray spectral ranges provide complementary information necessary to unravel outstanding puzzles concerning the physical mechanisms responsible for the winds. Ultraviolet P-Cygni line profiles indicate that the terminal velocities of the winds range from about 500 up to 3000 km/sec. Radio- and infrared-continuum emission allow one to infer the density profile of the wind. Recently, observations of x rays from early-type stars have suggested the existence of some previously unsuspected energy source in the winds that may be related to the anomalous ionization indicated by UV observations of O VI in the winds and to instabilities in the flow indicated by time variability in the UV line profiles.

The gross properties of the winds--mass-loss rate, terminal velocity, dependence on spectral type--can be understood in terms of a theory in which the wind is driven by radiation pressure due to resonance scattering by ions of trace elements in the wind. However, the theory has failed so far to explain details of the wind, such as ionization balance, velocity distribution, x-ray emission, and time variability. The theory is complex, involving coupled radiative transfer in lines, gas dynamics, and extensive atomic physics; its further development requires advances in each of these areas. It is also worth noting that the physical processes involved in the dynamics and radiative transfer in stellar winds are thought to be important in the emission-line regions of quasars and active galaxies. Since stellar winds are easier to observe than quasars, the development of the theory to explain the observed properties of the winds may be vital to understanding this outstanding problem of extragalactic astronomy.

The large mass loss in stellar winds has implications for many areas of astronomy. The most luminous early-type stars lose mass at a rate sufficient to alter drastically their evolutionary tracks on the H-R diagram. The stellar wind creates a cavity in the surrounding interstellar medium comparable with that created by a supernova explosion. In a dense interstellar cloud, the wind will modify the radio and infrared emission from the associated compact H II region, causing shocks and x-ray emission and perhaps affecting the formation and excitation of interstellar molecules. The stellar winds may play a major role in star formation, perhaps initiating this process by compressing gas and/or terminating it by disrupting the gas cloud. The huge cavities created by stellar winds in diffuse interstellar gas must be considered in the interpretation of the size and structure of the giant H II regions, both Galactic and extragalactic, and also in the interpretation of supernova shells, whose structure and evolution may be partly determined by the dynamical action of the stellar wind on the surrounding interstellar gas before the supernova eruption.

Several of the most luminous compact Galactic x-ray sources have, as binary companions, OB stars with strong stellar winds. The mass loss in the winds is sufficient to affect the orbital dynamics, perhaps greatly extending the duration of the x-ray luminous phase. In some cases, gravitational capture of the stellar wind may provide the mass transfer responsible for the x-ray emission. Even if the x-ray emission is not directly related to the stellar wind, the presence of an embedded binary x-ray source provides a unique probe of the structure and dynamics of the wind. The wind has sufficient column density to cause noticeable soft x-ray absorption that varies with orbital phase, providing a density probe. Observations of fluorescent x-ray emission lines from the wind are a likely possibility. The ionization of trace elements in the wind by the x-ray source has been observed indirectly through periodic variations of ultraviolet P-Cygni lines formed in the wind. The gravitational force and radiation from the x-ray source can substantially modify the dynamics of the winds, providing a probe of the physical mechanisms driving the wind.

Although we have learned a great deal about strong stellar winds in the past few years, our knowledge of this subject is still rudimentary. Radio observations with the Very Large Array (VLA), together with the greatly improved infrared observations expected from orbiting infrared telescopes, can provide a much better idea of the density structure of the winds of isolated OB stars and promise to reveal much about the dynamical effects of stellar winds of OB stars embedded in dense interstellar clouds. Ultraviolet spectroscopy has been severely constrained by the paucity of OB stars within the range of telescopes such as that on the Copernicus satellite, limiting our ability to derive systematic trends with stellar types. This situation should improve dramatically with the greatly improved sensitivities of the International Ultraviolet Explorer (IUE) and ST. However, a serious gap will remain in our ability to deduce the physical mechanisms of the stellar winds from UV spectra. We must develop an orbiting telescope capable of obtaining 912-1200-Å UV spectra, where many of the most physically revealing UV resonance lines are found. Further observations of x-ray sources associated with early-type stars will be invaluable for determining whether they are associated with compact objects in the system or arise instead from some unknown phenomenon. X-ray spectroscopic observations are critical; for most systems, these will require a second-generation x-ray telescope facility.

Finally, we emphasize that, despite impressive progress, gaps remain in the theoretical interpretation of the nature and implications of mass loss from early-type stars. Our rapidly developing capability to observe these phenomena creates an obligation to support an aggressive theoretical program.

2. Circumstellar Matter

Recent UV, infrared, and radio observations have confirmed the existence of extended circumstellar gas and dust shells around a variety of objects. Furthermore, most of these shell-bearing stars appear to be losing mass at very high rates. In some cases the mass loss rate may be as high as 10^{-4} solar masses per year. Knowledge of the chemical composition of these shells, as well as an understanding of the subsequent injection of this material into the interstellar medium, is crucial to unraveling the mystery of how the Galactic material is enriched in heavy elements and complex molecules. In some cases, the circumstellar material may be related to the formation of planetary and hence possibly life-supporting systems. A few notable examples of circumstellar-shell objects will serve to reveal the extent of the phenomenon.

Most very hot, early-type stars show evidence of high mass loss. Wolf-Rayet stars, Be stars, and Herbig-Haro objects exhibit excess infrared radiation owing to thermal free-free emission from hot plasmas and, in some cases, thermal emission from carbon or iron dust grains. UV and optical-line outflow velocities from Wolf-Rayet stars suggest mass loss rates in excess of 10^{-5}-10^{-4} solar masses per year. It has recently been shown that some Wolf-Rayet stars exhibit recurrent nova-like episodes in which dust shells composed of carbon or iron grains form and are driven into the interstellar medium. A concerted effort to obtain high-resolution infrared line spectra during these episodes may yield valuable information about the chemical composition of the processed material being returned to the interstellar medium.

Nearly all luminous late-type stars exhibit evidence of dense circumstellar dust shells composed of silicate or carbonaceous materials that are being steadily driven into the interstellar medium. These stars may, in fact, prove to be the greatest source of processed interstellar material. In the next decade, an effort should be mounted to investigate the detailed chemical composition of these shells as well as to determine more accurately the amount of mass being introduced into the interstellar medium from this source.

The Air Force Cambridge Research Laboratory infrared rocket survey has led to the discovery of a number of very cool (about 300 K) objects surrounded by dense graphite shells. These objects, probably in an advanced stage of evolution, may belong to a heretofore unknown population of stars that are spewing enormous amounts of carbon into the interstellar medium. Investigation of this possibility through observations by facilities such as IRAS and the Shuttle Infrared Telescope Facility (SIRTF) is of high priority for the next decade.

Finally, we have recently obtained new insight into the formation of, and ejection into, the interstellar medium of

dense carbon-rich dust shells during DQ Her-type nova erup-
tions. Again, studies using high-resolution spectral techniques may provide an accurate picture of the chemistry of this
processed material.

Radio observations of numerous late-type stars, Air Force
Geophysical Laboratory sources, and California Insitute of
Technology sources have shown that circumstellar dust and gas
shells often contain molecular masers. SiO, OH, and H_2O
masers appear to be present in many evolved stars and in protostellar candidates. Further studies of these masers using the
VLA and very-long-baseline-interferometric techniques should
spatially resolve some of them, thus providing new knowledge
about physical conditions in the circumstellar shells.

The resolution of as many infrared circumstellar shells as
possible using interferometric techniques is also of the
highest priority. This information is required if the shell
opacity, the shell mass, and shell mass-loss rate are to be
calculated to reasonable precision. Although IR interferometric measurements performed in the late 1970's have been very
promising, the available technology must undergo considerable
improvement if this information is to be obtained for a substantial number of objects.

F. Stellar Structure and Evolution

Significant progress in our understanding of stellar evolution
occurred during the 1970's. Among the most important advances
was the theoretical elucidation of late stages of nuclear
burning, including helium and shell flashes, carbon burning
stages in highly evolved massive stars, and the nucleosynthesis
of s-process elements. In addition, new insight was furnished
by theories of the pulsation process, which were expanded to
include effects such as nonradial oscillations and convection-
pulsation interactions. Finally, improvements in our understanding of the physical processes in stellar interiors and in
the calculation of theoretical isochrones now permit the dating
of star clusters of different chemical compositions, both in
the halo and in the disk, with sufficient accuracy to use them
as probes of Galactic evolution. The nagging problem of the
solar-neutrino discrepancy, however, remains a source of
concern in this field.

The next decade is likely to see a concerted attack on several key problems of stellar structure that have implications
for many branches of astronomy. Stellar winds and other physical mechanisms responsible for mass ejection from stars must be
better understood to assess the effect of mass loss on stellar evolution.

In the case of mass loss from late-type stars, detailed
studies of the Sun may provide useful information; a better

understanding of the convection mechanism, from both observational and theoretical points of view, is sorely needed. The discovery of magnetic activity cycles in late-type stars also offers new opportunities for research. Again, the Sun provides a unique opportunity for such studies.

Finally, many stars are known to rotate, but our understanding of the effects of rotation on stellar structure and evolution is still elementary.

1. Binary Systems

Some of the most fundamental questions regarding the structure and evolution of stars in close binary systems, whether degenerate or nondegenerate, still remain to be answered. These questions span the entire life cycle of binary systems, from birth through death. For example, the theory of formation of binary stars is in a chaotic state, and relevant observations are virtually nonexistent. We have observational studies of the bulk properties of binary systems but no direct understanding of the early stages of duplicity.

In addition to our relative ignorance of the formation stage, our understanding of the subsequent evolution of close binary stars is rudimentary. Accretion and/or mass outflow may play a vital role in the evolution of close binaries. In Galactic binary x-ray sources, the high luminosity generated by accretion enables direct observation of the accretion process, but for the vast majority of close binaries this is not the case. For many systems, current theoretical models predict the formation of thin disks confined to the system orbital plane, while both visible and ultraviolet spectroscopy indicate the opposite: the accreted material is so thick that the spectrum of the receiving star can be virtually completely obscured. Similarly, there is disagreement between theory and observation regarding the total ejection of material from binaries. Theory often predicts that mass is generally conserved in the system, while observations imply that much of the material may be lost.

There has been substantial progress in understanding cataclysmic variables and similar systems during the past few years, and many questions involving these systems appear to be on the threshold of solution. There is mounting evidence that nova outbursts are thermonuclear explosions, while dwarf nova outbursts appear to result from nuclear reactions caused by accretion events. Further theoretical studies combined with UV photometry and elemental abundance measurements could firmly estabish these models within the next few years. There is also observational evidence of an absence of systems with periods of a few hours--absence predicted by theory on the basis of evolutionary considerations. The evidence for this "gap" in periods should be subjected to rigorous observational tests over the next few years. The recent discovery of optical and x-ray pulses from cataclysmic variables and the compelling evidence for magnetic fields of some 10^6-10^8 gauss in these systems has raised a host of fascinating new questions, including the origin of these pulses and the mechanisms by which hard x rays are produced.

Despite a century of spectroscopy of binary systems, there is a surprising paucity of directly measured stellar masses. Here is one example of a crucial problem that has no gross technical barrier to its solution. The solution requires mainly that we assure ample facilities and observing time to interested astronomers to attack the problem vigorously in the next decade.

Our use of close binary systems as a probe of General Relativity is in its infancy, despite the fact that astronomical systems sometimes provide the only feasible tests of certain gravitational effects. The possibility that orbital period changes in the binary radio pulsar 1913+16 are due to energy dissipation through gravitational radiation and that the 164-day period in SS433 is a demonstration of relativistic Thirring-Lense precession are some of the few examples we can currently cite. We may expect the next decade to see a substantial increase in the number of possible test sites for General Relativity as improved observations identify more close binary systems with deep potential wells.

2. Objects of Unknown Nature

It seems almost gratuitous to state that serendipitously discovered, theoretically unpredicted phenomena will cause a significant fraction of the excitement in Galactic astronomy in the 1980's. Not only will such discoveries continue in the next decade, but a theoretical understanding of these phenomena will inevitably involve the interaction of theorists and observers from previously diverse specialties. What purpose does the serendipitous, unplannable discovery have in a planning document like this? Examination of some of the unexpected discoveries in Galactic astronomy during the 1970's illustrates vividly that such discoveries most often emerge on inauguration of new, innovative, higher-sensitivity observational facilities. Among many possible examples, three will be cited here to illustrate this point.

The rapid, aperiodic time variability of the x-ray emission from Cygnus X-1 and the existence of the entire class of x-ray burst sources were found only when previously well-known x-ray sources could be studied from satellites over time scales longer than was previously possible from sounding rockets. The understanding of these phenomena in terms of degenerate accreting stars has linked astronomers in such diverse fields as General Relativity and binary-star spectroscopy.

The existence of x-ray sources associated with globular star clusters was suspected from the first satellite observations, but only with the launch of the first satelliteborne modulation collimator experiments did high spatial resolution observations reveal that the sources are located very near the centers of the clusters. Understanding these observations has required close interaction between the x-ray astronomer and the stellar dynamicist; the true nature of these objects is still in dispute.

The discovery of the high-velocity, periodic red shifts and blue shifts in SS433 involved observation of a star cataloged and forgotten two decades previously but noticed again as a result of modern x-ray and radio interferometric observations. Ironically, the almost-forgotten SS433 may prove to be an entirely new class of Galactic objects. If the currently popular models of an ejected, collimated jet of relativistic gas prove correct for this object, a link will be forged between studies of the late stages of stellar evolution and theories of twin-lobed extragalactic radio sources.

It would be an omission not to include these "objects of unknown nature" in a list of future problems of Galactic astronomy in the 1980's. Recent astronomical history such as that cited above shows that such important surprises will emerge if we maintain vigorous and expanding support of appropriate facilities. These discoveries will be exciting bonuses to the more predictable scientific returns of these investments.

G. Stellar Deaths

While considerable progress has been made in the last decade in our understanding of the way in which stars die, many key questions remain unanswered. The study of the fates of highly evolved stars makes contact with many other currently active areas of astrophysics and is likely to be one of the most important research topics of the 1980's. Among the most important questions are: Which stars collapse, and which have more quiet deaths? How do degenerate dwarfs, once formed, evolve? Of the stars that suffer gravitational collapse, which form neutron stars, and which black holes? Do black holes exist? Can highly evolved stars collapse, forming neutron stars or black holes, without creating a highly visible supernova explosion? What ultimately becomes of rotating neutron stars in close binary systems?

1. The Fate of Intermediate- and High-Mass Stars

Recent progress in detecting and measuring mass loss from evolving stars has provided evidence that a substantial fraction of a star's mass may be lost during its hydrostatic evolution, so that stars with initial main-sequence masses as large as 8-10 solar masses may form degenerate dwarfs rather than undergo gravitational collapse. Knowledge of mass loss is critical both for determining the correct initial models to use in detailed studies of stellar collapse and for comparing the numbers of observed condensed remnants (such as degenerate dwarfs, neutron stars, and black holes) with theoretical predictions. Infrared, optical UV, and perhaps x-ray studies of stellar mass loss, combined with optical and UV studies of supernova outbursts in other galaxies and the theoretical modeling of mass-loss processes, may be able to provide this knowledge during the next decade.

2. Supernovae and Their Remnants

Understanding supernovae is central to progress in many areas of astrophysics. Supernovae are thought to be responsible for the formation of both neutron stars and black holes. Most of the heavy elements in the interstellar gas and in second and subsequent generations of stars are currently thought to have been formed in the interiors of supernovae. Supernovae are possibly the single most important energy source for the interstellar medium and for Galactic cosmic rays. The remnants of supernovae provide nearby laboratories for studying the physical processes that govern extragalactic phenomena: the Crab nebula can provide insights into nonthermal processes occurring in quasars and active galactic nuclei, whereas study of the interaction between cool clouds and hot, x-ray emitting gas and Cas A can add to our understanding of the behavior of the gas in galaxies located in x-ray emitting clusters of galaxies. Finally, because of their great luminosity, supernovae hold considerable promise as calibrators of the extragalactic distance scale.

Supernovae are sufficiently rare that in modern times the actual supernova explosion has been studied only for extragalactic supernovae. The last supernova observed in our Galaxy was Kepler's supernova of 1604. Studies of supernovae in other galaxies and observations of pulsars and supernova remnants in our own Galaxy suggest that the supernova rate in our Galaxy is of order 1 every 10-20 years. The many supernovae that have occurred in our Galaxy since 1604 have presumably been so highly obscured that they have escaped optical detection (only one, Cas A, has been discovered subsequently). Galactic supernovae should be more readily detectable in the infrared and x-ray regions of the spectrum, and astronomers should be prepared to devote intensive study to one, should it occur.

By contrast, the remnants of supernovae live for thousands of years, and many are available for study in our Galaxy. Major questions about these objects that should be addressed in the coming decade are: Where have all the remnants gone?

Radio, infrared, and x-ray observations in this decade should allow the discovery of other young supernova remnants besides the historically observed ones and Cas A. How is the nature of the remnant related to the nature of the supernova? Supernovae are classified primarily as Type I, which are thought to come from relatively old stars, and Type II, which originate from massive, early-type stars; the study of young remnants should tell us much about the supernovae themselves. How are heavy elements actually injected into the interstellar medium—as gas or dust, in cool clouds or in hot diffuse gas—and how are these elements dispersed? Here again, the study of young remnants such as Cas A is important; for example, optical spectra of Cas A show relatively cool, high-velocity filaments that consist essentially of heavy elements, whereas x-ray observations of hot gas reveal only a modest enhancement in metal abundances. What are the dynamical consequences of the expansion of a supernova remnant into an inhomogeneous medium—how large does an old remnant become, how do remnants affect molecular clouds, and in what cases can supernova remnants trigger star formation? Finally, what are the physical processes involved in supernova remnants such as the Crab nebula, with an embedded pulsar, or W50 with SS433, and how are these related to quasars and active galactic nuclei?

All these problems are ripe for solution in the 1980's. Observationally, spectroscopy of faint gas in a number of spectral regions is required: radio and infrared observations of molecules in supernova remnants; infrared observations of nebular emission lines in obscured remnants; and optical, UV, and x-ray observations of closer remnants to unravel the complex physical processes occurring there. Continuum observations in different spectral regions are useful for discovering new remnants and for studying nonthermal emission. On the theoretical side, increasingly sophisticated treatments of the hydrodynamics of the expansion of remnants and of the radiation they emit are required.

3. Formation of Neutron Stars and Black Holes

One of the most important unsolved problems in astrophysics is the process by which neutron stars and black holes are formed. During the 1970's, reliable treatments of the interaction of neutrinos with dense matter, of the equation of state of hot matter at densities up to nuclear density, and of the hydrodynamics of spherically symmetric gravitational collapse have been achieved. Nevertheless, many other aspects of the problem are still poorly understood.

Studies of the collapse of iron-core stars (star with initial mass greater than some 10 solar masses) have now reached the point where a sustained theoretical effort may lead to a solution during the next decade. The dynamical behavior of such stars during their initial collapse and first rebound is now understood, and there is also widespread agreement on the next problems to be addressed. These include the effects of large-scale mixing due to convection, the effects of moderate rotation, and the amount of gravitational radiation emitted during collapse. Careful fluid-dynamical studies and multi-dimensional hydrodynamic calculations will be needed to solve many of these problems. It may also be necessary to develop accurate treatments of the equation of state of hot matter above nuclear density and of the dynamical effects of rotation.

The fate of intermediate-mass stars (in the range 4-10 solar masses) is at present uncertain. Support should be continued for theoretical and observational studies of the late stages of evolution of stars in this mass range, and for theoretical studies of detonation, deflagration, cooling, and heating under conditions similar to those in the cores of these stars.

The apparent conflict between the number of pulsars in the Galaxy and estimates of supernova rates is also an important problem. The significance of this conflict can be clarified by progress on the issues just described and by study of "nonstandard" collapse scenarios, which might produce neutron stars without the copious optical emission of classical supernovae.

H. Compact Objects

1. Black Holes

The first observational hints of the existence of black holes in the Galaxy emerged during the 1970's. Current evidence is, however, extremely scanty. One x-ray binary system has an as yet unseen component that is inferred indirectly to be compact and more massive than theoretically calculated limits for white dwarfs and neutron stars. Several x-ray systems show rapid, aperiodic time variability, indicating that the emission probably originates in a compact region of size appropriate to a neutron star or smaller. In addition, the spatial distribution of a handful of globular-cluster x-ray sources with respect to the cluster visible-light centers suggests that these sources may be more massive than the most massive objects expected on the basis of current theories of stellar evolution.

Clearly, the evidence is tantalizing but unsatisfactory, and in the next decade we may expect a variety of attempts to verify the existence of black holes. The temporal behavior of binary x-ray sources requires continued study with large-area detectors; the more recent generation of modest-area imaging sensors is not suitable for this work. Current capabilities of radial-velocity spectroscopy are just barely sufficient to

derive mass functions for many of the ineresting binaries, so
we may hope for substantially improved and increased observations in this area. The mass and light distributions of the
cores of compact globular clusters are most amenable to study
by satelliteborne high-spatial-resolution imaging and spectroscopic experiments, and we may expect the first of these
results in the next decade. Finally, substantial theoretical
advances are still required to aid in defining possibly unique
observational signatures of black holes.

2. White Dwarfs

Although white dwarfs are the supposed endpoint of evolution
for the majority of stars, we know surprisingly little of their
evolution and properties. We have no observational knowledge
of the evolutionary transition between the planetary-nebula and
white-dwarf phases in low-mass stars. Observations during the
next decade of hot sdO and sdB stars, especially in the far
ultraviolet, may substantially clarify the properties and space
densities of these stars. These data in turn should lead to
theoretical advances such as an understanding of the
evolutionary importance of neutrino cooling.

Both extreme ends of the white-dwarf luminosity function
provide interesting challenges for the 1980's. Recently,
optical surveys seem to indicate that there are very few faint
white dwarfs in the Galaxy, contrary to the numbers expected
from formation at a constant rate. Accurate and reliable
theoretical calculations of white-dwarf cooling are in process
and will become available during the next few years. These
theoretical models, when combined with detailed observations of
individual stars from the optical surveys, can show whether the
absence of faint white dwarfs is due to significant changes in
the white-dwarf birth rate during the life of the Galaxy (as
now seems likely) or to errors in the present theory of white-
dwarf cooling.

The hotter end of the white-dwarf luminosity function is
currently quite uncertain. These objects radiate primarily in
the ultraviolet and extreme ultraviolet regions, where surveys
for such stars in the immediate solar neighborhood have barely
begun. An additional outstanding unsolved problem in the
evolution of white dwarfs is the nature of and relationship
between the DA, DB, DC, and 4680 white dwarfs. This problem
may be ripe for solution during the 1980's. Recent studies
suggest that gravitational settling plays a role, but
additional theoretical work is needed to determine reliably the
effects of convection, rotation, binary membership, and other
factors.

Finally, there are still only a handful of dynamically
measured white-dwarf masses, a situation that it is hoped can
be improved in the next decade.

3. Neutron Stars

Demonstration of the existence of neutron stars is one of the
major astrophysical breakthroughs of the 1970's. These objects
have masses comparable with the Sun's, but their densities
range up to more than 10^{14} times that of normal matter. Hard
x-ray spectral observations imply that magnetic fields associated with neutron stars may be as high as 10^{12} gauss.

The first neutron stars were discovered as radio pulsars,
some of which have been shown to be in binary systems. Four,
including the Crab and Vela pulsars, were subsequently detected
at gamma-ray energies. A second group of neutron stars was
discovered as pulsating x-ray sources. All of these appear to
be members of close binary systems, where the x-ray emission
seems to be a consequence of mass transfer to a magnetic neutron star.

It is apparent from all of these remarkable properties that
neutron stars offer a unique testing ground for our understanding of the fundamental physical laws of nature. There are many
important questions regarding the nature of neutron stars that
must be answered before the physical nature of these objects is
well understood. A substantial body of high-quality spectral
information about neutron stars is already available, and more
is likely to be forthcoming over the next few years. The main
factor impeding progress in this area is the absence of reliable models, a situation that points to the need for a substantial theoretical effort on these problems during the next
decade.

Major neutron-star problems that appear to be ripe for
study during the 1980's include determination of the mass range
for neutron stars; elucidation of the relationship between
neutron stars and black holes; the development of observational
tests for the equations of state of neutron degenerate matter
and for Einstein's General Theory of Relativity; and the
development of theories to describe the internal structure of
neutron stars. Further studies are also required into the
nature of the magnetic properties and radiation mechanisms of
pulsars. In addition, little is now known of pulsars, neutron
stars, and black holes as stellar populations in the Galaxy.

Solutions to these problems will depend in part on the
availability of key x-ray timing and faint-object proper-motion
measurements. This will require x-ray satellites with high
sensitivity and good background rejection, such as can be
achieved with focusing detectors.

4. Binary Pulsars

The General Theory of Relativity predicts that two masses
orbiting about one another will emit gravitational radiation.
Until recently, this major prediction of the theory had never

been tested. However, an opportunity for such a test presented itself with the discovery in 1974 that the pulsar PSR 1913+16 is a member of a close binary system. Two more binary pulsars have been discovered since 1974. Of more than 300 pulsars discovered to date, only these three have been found to be members of binary systems. PSR 1913+16 orbits its companion every 7.75 hours. General Relativity predicts that the system should radiate about 10^{33} ergs/sec in gravitational waves, comparable with the visible-light output of the Sun. As this energy is lost, the two stars should slowly spiral inward, the binary period decreasing at the rate of 76 µsec per year if both members of the system are neutron stars. By comparison, the arrival time of individual radio pulses can be established to within 50 µsec. By 1979, arrival-time measurements showed that the orbital period is decreasing at a rate that is consistent with the prediction of General Relativity. This is the first test of General Relativity ever made on objects outside the solar system and is striking evidence for the existence of the quadrupole gravitational radiation predicted by the theory. The measured decrease in the orbital period appears to rule out several other theories of gravitation.

The same arrival time measurements show that the orbit of PSR 1913+16 precesses by 4.226 $\pm$ 0.002 deg/year. Assuming that this is entirely due to General Relativistic effects, it implies that the masses of the pulsar and its companion are 1.39 $\pm$ 0.15 and 1.44 $\pm$ 0.15 solar masses, respectively. If confirmed, this would represent the first direct determination of the mass of a pulsar.

During the next decade, further observations of PSR 1913+16 should pin down the system parameters with even greater accuracy, resolve ambiguities that now exist, and test General Relativity with even greater precision. Because of the fundamental importance of measurements of this type, the possibility of discovery of other binary pulsars is by itself sufficient justification for further pulsar searches.

5. Gamma-Ray Sources

In the next decade we may expect a substantial advance in the observation of point gamma-ray sources, many or most of which are presumably compact Galactic objects. At the moment there are only a dozen or so such sources, all poorly located and detected with modest statistical precision. The next decade will, it is hoped, see advances similar to those of the early 1970's in x-ray astronomy: increases in the total number of known sources, the detected signal-to-noise of each source, and the source positional accuracy. The last is vitally necessary to permit optical identifications of counterparts. At least crude spectral information will be necessary to shed light on some production mechanisms and also to fulfill the exciting

prospect of observing gamma-ray lines from nuclear processes
that almost surely occur in selected astrophysical environments.
Finally, the nature of the gamma-ray burst sources is still
enigmatic and quite possibly fundamentally different from that
of the steady sources. These sources offer a good prospect of
yielding precise positional data via simultaneous observations
by multiple detectors over interplanetary baselines, and many
expanded opportunities for such observations may arise in
connection with planetary programs of the 1980's.

VI. FUNDAMENTAL PROBLEMS AND NEEDS FOR THE 1980'S

A. Fundamental Astronomical Problems

The formulation of many of the fundamental problems of Galactic
astronomy has changed little over a long period of time.
Although astronomers continue to use new techniques as
opportunities arise, astronomical research in many areas shows
a continuity over decades that is not shared by many other
sciences. Thus, the problem of distance measurement still
requires improved calibration of the absolute magnitudes of
variable stars and extensive long-term programs in astrometry;
the determinations of the Galactic luminosity function entails
painstaking measurements of great numbers of low-luminosity
stars; and specifying the hierarchy of structures in the interstellar medium requires large quantities of radio data whose
collection will continue to involve astronomers for years.
   The Working Group believes that the continued good health
of Galactic astronomy requires that adequate attention and
support be given to long-term observational programs.

B. Basic Astrophysical Data

To ensure the continued progress of Galactic astrophysics
during the next decade, renewed effort will also be needed in
the measurement and in the calculation of basic nuclear and
atomic data. More accurate cross sections necessary for
investigations of nucleosynthesis and stellar evolution are now
required. Opacities for stellar interiors and atmospheres
reflecting a wide variety of chemical mixtures must be calculated. Particularly important are the opacities due to
molecules found in stellar atmospheres and envelopes; such
molecules determine the structure of the outer layers of cool
stars and especially of red giant stars, which in turn dominate
the light we receive from elliptical galaxies. Laboratory
f-values, as well as the cross sections for collisional processes, are essential for the determination of the physical
conditions and abundances in both stellar atmospheres and the

interstellar medium. In radio astronomy, measurements of molecular microwave spectra are particularly important for studies of the interstellar medium.

C. Fundamental Physical Problems in Stars

In the next few years, special attention needs to be given to fundamental problems in stellar physics, such as the properties of thermal convection, the interaction of convection with radiation fields, and investigations of nonstationary phenomena in stars. It is clear that, in addition to the frontier areas of star formation and of stellar deaths singled out in this chapter, much research effort will concentrate on such problems and internal mixing and mass loss from stars during the middle phases of stellar evolution. These processes, which bear directly on the broad problem of the chemical evolution of the Galaxy, are barely understood at this time. Much of the theoretical work in this area will be computational and will rely on numerical techniques that are still largely experimental. The Working Group emphasizes the importance to Galactic astronomy of research in basic stellar physics and the necessity of providing workers in this field with easy access to versatile interactive computing facilities.

D. Observing Needs for the 1980's in Galactic Astronomy

One of the most characteristic features of Galactic astronomy is that it typically progresses by detailed study of a wide variety of individual objects. In addition, the solutions to many of the fundamental problems in Galactic astronomy will require long-term observational programs and the availability of facilities that can respond in a timely and effective manner to transient phenomena.

These requirements have important implications for the strategy that should be adopted for the development of Galactic astronomy in the 1980's and beyond. Probably no single factor has restricted the development of Galactic astronomy during the last decade as much as has the lack of adequate amounts of large-telescope observing time. Furthermore, a tendency for large observatory facilities to be rather inflexible in terms of scheduling has restricted our ability to respond to and obtain adequate coverage of transient phenomena. This suggests that we explore the advantage of building a number of standardized ground-based thin-mirror telescopes of about 3-m aperture. Such instruments may turn out to be extremely cost-effective photon collectors. These telescopes should be deployed so that they can be effective in responding to long-term programs and studies of transient phenomena.

Relocation of several regional or university-based facilities might be beneficial in this regard.

It is worth recalling that the Greenstein report (<u>Astronomy and Astrophysics for the 1970's, Volume 1: Report of the Astronomy Survey Committee,</u> National Academy of Sciences, Washington, D.C., 1972, page 117) states that "Many of the scientific goals for the coming decade will involve very faint objects and require extended observational programs. This style of operation may thus become more common (for at least part of the available time) at national observatories." Experience shows that this hope was not realized during the 1970's.

The Working Group therefore wishes to emphasize the need for continued support for the activities discussed above. While recognizing that these basic and fundamental scientific studies are no more timely for the 1980's than for other decades, we believe that they provide the foundation for the knowledge gained from Galactic and all other branches of astronomy.

4

Extragalactic Astronomy

I. INTRODUCTION

In writing this chapter, we took our charge to be twofold:
first, to review the significant progress made in extragalactic
astronomy since the writing of the Greenstein report in 1970;
then, to identify those scientific goals that seem to be most
critical or fruitful for the continued progress of extragalactic astronomy during the period 1980-1990, to rank-order or
otherwise estimate the relative importance of these goals, and
to suggest means by which these goals might best be achieved
during the coming decade.

In keeping with the emphasis of this charge, our final
summary is a list of scientific aims rather than a rank-ordered
list of facilities for the next decade. In many cases, we felt
doubtful about the best technical means to achieve a particular
goal. In other cases, we felt uninformed about the design
capabilities of facilities that have already been proposed. In
short, it did not seem appropriate for us to debate the merits
of competing facilities in this chapter, which is intended
primarily as a scientific review. Our references to facilities
therefore usually do not cite particular projects but instead
refer only to specific wavelength ranges, spectral resolutions,
and sensitivity limits. We did, however, rely on our overall
knowledge of current project planning as a general guide to
what will be technically feasible over the next decade.

The only exceptions to this policy arose in connection with
a few specific projects that have already been funded or are
well advanced in the agency approval process. We felt that it
would do no harm to remind the astronomical community of the

great advances to be expected from these facilities, which are deserving of continued fiscal support to ensure their maximum productivity.

The third, fourth, and fifth sections of this chapter present a fairly comprehensive review and prospectus for each of five broad areas of extragalactic astronomy; these sections are intended to be reasonably complete but not exhaustive. Many valuable projects have been omitted for lack of space, particularly if they do not require a notable rearrangement of priorities or reallocation of resources. Our emphasis has been on programs that either demand a significant commitment of new resources or deserve a larger share of existing resources.

After these summaries were assembled, we pooled the major programs from all areas and voted on their relative importance. Several programs emerged as clear favorites. With the help of this straw vote, we were able to formulate 15 scientific objectives for the next decade that encompassed all of our favorite programs. A second vote then produced a final ranking among these 15 questions, which is presented in the next section together with a brief commentary on each entry; this section thus summarizes the results of our deliberations in a compact and easily accessible form. To obtain a complete picture of our conclusions, the reader should also consult the last section, which contains remarks of a general nature that did not fit naturally into any preceding sections.

The final rank-ordered list of priorities is intended not as a list of the 15 questions of ultimate importance in extragalactic astronomy but rather as a realistic set of goals for the next decade. Many of the obvious "ultimate" questions are not included in our list simply because we could think of no good way to answer them yet, or because we cannot realistically expect firm answers within the next 10 years. An example is the determination of q_0, the deceleration parameter for the Universe. It is by now very clear that the measurement of q_0 using lookback studies of galaxies, clusters of galaxies, and other objects is plagued by problems of source evolution-- hence our decision to emphasize evolutionary studies of these objects during the 1980's. Depending on the outcome of this preliminary groundwork, the measurement of q_0 via lookback studies could become a high-priority scientific goal further downstream.

We recognize a potential failing of this chapter--its conservatism. Our final 15 ideas certainly constitute good science, but they are largely the ideas that leap immediately to the mind of anyone working in extragalactic astronomy. Perhaps we can be excused for our lack of daring on the grounds that we had to cover an enormous amount of material. More importantly, this chapter truly reflects the consensus of the whole panel, and convolution by consensus eliminates the valleys only at the cost of leveling the peaks.

On the other hand, we note that the spectacular progress made in extragalactic astronomy during the last decade arose from just such a balanced, flexible program as we propose here--a program that exploited new wavelength regions, spectral resolutions, and sensitivity limits in systematic fashion. As we all know, many of the anticipated gains were in fact achieved, but, just as importantly, totally unexpected new phenomena were also discovered. We anticipate that equally exciting advances will be made during the 1980's.

II. THE MOST IMPORTANT QUESTIONS IN EXTRAGALACTIC ASTRONOMY FOR THE 1980'S

In this section we present our list of the most important questions in extragalactic astronomy for the next decade. To convey a sense of our priorities, the list is rank-ordered, topics near the end receiving roughly half the votes of those at the top. This means that the topics appear in a somewhat illogical order, but the information conveyed by the rank-ordering was deemed to outweigh this disadvantage.

1. What is the structure of the central energy source in quasars and active nuclei, and what fundamental physical processes are involved in the energy release?

Quasars are the most luminous objects known in the entire Universe. Some quasars emit thousands of times as much energy as a normal galaxy, yet their dimensions appear to be incredibly tiny by astronomical standards--no larger than our own solar system. Current theories suggest that the most intense gravitational fields in the Universe--apart from those associated with the big bang itself--are to be found in the cores of quasars. A clear understanding of quasars may even ultimately serve as a test of the General Theory of Relativity and alternative theories of gravity.

Progress in modeling quasars requires answers to a number of related questions: (a) Is the current working model-- involving energy release through accretion of matter onto a central massive collapsed object--correct? (b) What is the surrounding structure of matter and magnetic fields? (c) How are objects within the broad family of "active galaxies" related to one another (e.g., radio-loud and radio-quiet quasi-stellar objects, optically violent variables, BL Lac objects, radio galaxies, and Seyfert nuclei)? (d) What are the processes producing the emitted radiation, particularly in the gamma-ray and x-ray regions?

Emission processes in quasars are best studied through analysis of the spectral energy distribution and polarization of the radiation as a function of frequency; observations in the far-infrared, x-ray, and gamma-ray regions are especially important.

Increased angular resolution is crucial to delineate source structure on the smallest possible scales. Detection of rapid variability will probably remain our strongest constraint on source size but carries almost no direct information about the actual geometry; direct imaging through interferometry will therefore be of high importance, too. Space very-long-baseline-interferometry (VLBI) techniques employing baselines substantially larger than 10^4 km would test the 10^{12} K brightness limit for synchrotron emission and map the radio structure on exceedingly small scales. Ground-based VLBI techniques now enable us to resolve structures down to 10^{-4} arcsec, corresponding to dimensions of 1 pc for quasars at moderate distances. If comparable spatial information in the visible and UV regions were available, we could study the structure and dynamics of the emission-line region directly, and this information would be crucial in answering points (b) and (c) above. We therefore urge the development of optical interferometric techniques ultimately leading to resolutions of 10^{-4} arcsec, from Earth orbit if necessary.

2. What do distant galaxies look like beyond red shifts of 0.5?

When we look at distant, highly red-shifted objects, we are also looking far back in time. This fact gives astronomers the unique ability to see galaxies exactly as they were many billions of years ago. During the 1980's we will pursue this technique further to follow up on the recently discovered hints of dramatic galaxy evolution at large lookback times. Important clues to the epoch of initial star formation in galaxies should emerge.

Space Telescope (ST) will provide high-resolution photographs indispensable to the morphological study of distant galaxies. Broadband energy distributions from 1500 Å to 5 µm in the rest frames of near and distant galaxies will be sensitive to the aging of stellar populations. Deep galaxy counts and red-shift surveys to a visual magnitude of 23 are vital to the measurement of the luminosity function, rates of evolution, and clustering properties of high-red-shift galaxies. Additional facilities essential to this program include larger ground-based optical telescopes; multiplexing red-shift detectors; improved infrared sensitivity out to 5 µm; and high-efficiency, two-dimensional detectors in the wavelength range 1500 Å to 5 µm.

Detection of primeval galaxies actually undergoing collapse or coalescence will remain an important observational goal for

the 1980's. The spectral distribution of the emitted radiation
from these objects is hard to predict, but searches may concentrate on the 1-keV x-ray radiation from an early burst of
supernovae or on copious infrared and microwave emission from
the stars themselves and from radiation by enveloping dust
clouds. A large, orbiting, infrared telescope diffraction-
limited at wavelengths longer than 2 μm would be the ultimate
instrument of choice for infrared observations.

3. <u>What is the amount, extent, and nature of the mass that</u>
<u>surrounds galaxies?</u>

We now have strong evidence that individual galaxies are
frequently immersed within massive "halos" or "coronas" consisting of dark matter quite unlike the ordinary stars and gas
with which we are familiar. The discovery of these major
unseen constituents merits intensive investigation in the
1980's through observations aimed at detecting any faint halo
material. We must also develop new dynamical tests for unseen
matter, such as the measurement of radial velocities of globular clusters in other galaxies and the temperature and density
profiles of galactic x-ray halos. Detailed studies of interactions with orbiting gas or companion galaxies may also be
used to probe the extent and shapes of these puzzling mass
distributions. Theoretically we must consider how these halos
arose in the first place, what the possible forms of matter are
in the halos, and whether the halos might not even predate
visible galaxies.
    Essential facilities here include the enormously strength-
ened 21-cm capability of the Very Large Array (VLA), x-ray
telescopes in space, and, for the globular clusters, improved
red-shift technology involving larger optical telescopes,
multiplexing red-shift detectors, and high-efficiency
two-dimensional detectors. Adequate theoretical support is
also vital.

4. <u>What is the amount, extent, and nature of the mass in</u>
<u>groups and clusters of galaxies?</u>

The mass we can see in individual galaxies is only 5 to 10
percent of the total needed to stabilize most groups and clusters of galaxies. Unbiased samples of galaxy red shifts at z
less than 0.1 with errors of less than 75 km/sec are needed to
test whether groups and clusters are stable, to provide reliable estimates of velocity dispersions and streaming motions in
the Hubble flow, and to trace the mass distribution within
clusters and groups. Generous time allocations on optical
telescopes of moderate size are required, together with multiplexing red-shift detectors.

5. How has the clustering of galaxies evolved with time?

It is now possible to observe the clustering of galaxies out to
a red shift of 1 or 2 and to see how it differs from
present-day clustering. The results of such studies will
profoundly influence ideas on the origin of structure in the
Universe. Red-shift samples to z = 1 with an accuracy in z of
about 10 percent are needed for precise measures of galaxy
clustering and for calibration of the galaxy luminosity
function. Red-shift errors equal to or less than 100 km/sec
will be needed for studies of cluster dynamics at large z.
(Reducing the errors to this amount will be exceedingly
difficult but should be kept in mind as an ultimate goal.)
Samples of the angular distribution of galaxies together with
broadband colors will yield a cruder yet essential picture of
the clustering at red shift of 1 or 2, K magnitudes of about 18
or 20, and V magnitudes of 23-25.

Essential facilities here include ST, larger ground-based
optical telescopes, and improved red-shift technology as noted
above. A dedicated N-body parallel-processor computer (with N
in the range 100,000 to 1,000,000) could make a uniquely
important contribution to the development of theoretical ideas.

6. Do the nuclei of normal galaxies contain dormant quasi-stellar objects?

Low-level activity is found in the nuclei of many elliptical
and spiral galaxies, including our own. Is there a connection
between this activity and the more spectacular outbursts in
radio galaxies, Seyfert nuclei, and quasars? Is it true that
many, or even all, galaxies harbor in their nuclei massive,
collapsed objects that once were the seat of violent activity
but that are now "extinct" or merely "dormant"?

Very-high-resolution radio maps from the VLA and from VLBI
will be our most important radio tool in this area in the next
decade. In the near future, ST's will greatly increase spatial
resolution for optical imaging of nearby galactic nuclei.
Ultimately, however, optical interferometric techniques should
be developed to utilize fully the potential resolving power at
optical wavelengths. ST badly needs a spectroscopic detector
that will allow dynamical studies to be made at high angular
resolution deep within the cores of nearby galactic nuclei.
X-ray telescopes can also be used to probe both active and
normal galactic nuclei. Luminosities and variations in x-ray
intensity and spectral distribution should prove to be
sensitive indicators of both nuclear size and mass.

7. <u>Are there major astrophysical phenomena yet to be</u>
<u>discovered?</u>

If the past decade is any guide, important and totally
unexpected phenomena will be discovered whenever new wavelength
regions are opened up or when sensitivity in existing bands is
significantly increased. The prime unexplored frequency ranges
at present lie in the far infrared, at gamma-ray wavelengths,
and at radio wavelengths of 10 MHz or less. Observations at
these wavelengths can only be made above the Earth's atmosphere. Significant sensitivity gains seem possible in all
other bands as well, most notably in x rays. Neutrinos,
gravitational waves, and high-energy cosmic rays are additional
untapped sources of information.

8. <u>What is the relationship of quasi-stallar objects and</u>
<u>active nuclei to the types of galaxies and galaxy groups in</u>
<u>which they are embedded?</u>

The first goal here must be to test the basic hypothesis that
quasi-stellar objects (QSO's) are active nuclei of galaxies.
If the answer is affirmative, the relationship between the
properties of an active nucleus and the morphological type and
mass of its associated galaxy will tell us much about the
conditions necessary for quasar activity. The role of the
environment can be explored by studying the numbers and types
of neighboring galaxies.
      ST is the single most important instrument for studying the
nature of the associated galaxies. Both ST and larger ground-
based optical telescopes can make important contributions to
investigations of cluster membership.

9. <u>What are the fundamental structural parameters of</u>
<u>galaxies, how do they vary from one galaxy to another, how do</u>
<u>they depend on environment, and what do these variations tell</u>
<u>us about galaxy formation and evolution?</u>

We are still ignorant of many of the most basic structural
parameters of normal galaxies. Accurate surface-brightness
distributions exist for only a handful of objects and are
virtually nonexistent for barred galaxies, for example. These
and many similar fundamental measurements are lacking because
systematic surveys of basic properties of galaxies are still
incomplete; this lack is particularly distressing because
correlations among the few basic parameters we do know about
provide some of our best constraints on theories of galaxy
formation and evolution.

For example, one of the major developments of the past
decade has been the growing suspicion that galaxies may not
evolve as closed systems in isolation but may instead be
profoundly affected by interactions with their environment.
Stripping of gas by an external intergalactic medium, tidal
truncation, cannibalism, and mergers are some of the mechanisms
that have been suggested. The importance of external influences
like these is best studied through careful intercomparison of
basic galaxy parameters as a functon of environment. Without
careful standardized measurements of many dozens of galaxies,
such trends cannot be examined.

The facilities needed for this wide-ranging program are
many and diverse. Most important is a thorough knowledge of
the nearest, most easily studied galaxies. Much of this work
can be done with moderate-to-large-sized optical and radio
telescopes; it is time-consuming, however, and at the moment is
severely limited by adequate access to facilities, especially
in the optical range. Development of efficient two-dimensional
detectors from 0.3 to 2.5 µm would be immensely helpful
here. We also need systematic surveys of the x-ray properties
of galaxies, which will tell us much about their hot-gas
content. VLA maps of neutral-gas distributions in cluster and
field galaxies are likewise important. ST will provide
high-angular-resolution maps of brightness profiles in galaxy
cores.

Unique additional information will come from lookback
studies of distant galaxies, on which environmental influences
could be quite different owing to a different degree of clustering at earlier epochs. For this purpose, the high angular
resolution of ST, the increased spectroscopic capability of
larger ground-based optical telescopes, and x-ray imaging of
distant clusters are all essential.

10. What is the history of the Universe at very large red shifts, as deduced from the spectral distortions and anisotropy of the microwave background?

The cosmic microwave background radiation provides a uniquely
powerful probe of the early Universe out to a red shift of 1000
or more. It may even reveal density fluctuations in the early
Universe from which the present-day large-scale structure
developed. Detection of fine-scale anisotropy on scales of 1
arcmin or less will therefore provide a critical test of the
entire gravitational-instability picture of galaxy formation
and clustering and of the effects of the clumpy matter
distribution along the line of sight. Measurements such as
these depend on improvements in millimeter-wave technology
generally and may ultimately require a large (30-m) radio
antenna in Earth orbit.

It is of crucial importance to confirm the spectral distortions of the cosmic blackbody radiation spectrum recently claimed to exist at the 10 percent level. These distortions could have a variety of causes, including a pregalactic generation of stars. Spatial variations in the spectral distortions may thus provide vital information on the early evolution of protogalaxies and on the origin of heavy elements and dust grains. Angular resolution on a scale of 1-60 arcmin is required at millimeter and submillimeter wavelengths.

The large-scale anisotropy provides an important test of the standard big bang cosmology with regard to large-scale shear, rotation, and inhomogeneities. For these large-scale variations, measurements from 3 mm to 1.5 cm on angular scales of 10° or so and with a sensitivity of 0.1 mK are required; these capabilities will be provided by the Cosmic Background Explorer (COBE) satellite.

11. What is the origin and evolution of the intracluster medium?

The intracluster medium has probably had a major impact on galactic evolution, and vice versa. X-ray imaging with higher sensitivity and at energies extending to at least 7 keV is required to determine the gravitational potential, gas content, and iron abundance for individual clusters. Observations over a range of red shifts will allow us to follow the evolution of these quantitites in time.

Given the gas temperature from x-ray observations, maps of the microwave diminution for the same clusters will tell us the column density of electrons along the line of sight. (This may require a space experiment, e.g., the 30-m radio telescope mentioned earlier working at wavelengths of 1 cm or less.) X-ray and microwave observations can then be combined to determine absolute distances.

Low-frequency radio observations below 50 MHz and high-energy x-ray spectral scans (and, if possible, imaging as well) from 10 to 100 keV are needed to study the relativisitic-particle distribution in the clusters. Lookback studies with x-ray telescopes, ST, and the largest possible ground-based optical telescopes can elucidate interrelationships between galaxy evolution, galaxy clustering, and the intergalactic medium.

12. How have the life histories of stars affected the evolution of galaxies, and vice versa?

A knowledge of the physics of star formation and evolution is vital to our understanding of galaxy evolution as a whole. The lack of a well-founded theory of star formation is still the most serious problem in this area and continues to block any a priori theory for galaxy formation. Although the major advances in this area must be furnished by theoretical work, star-forming regions in our own and nearby galaxies can provide important empirical evidence on the physics of star formation and the initial mass function. Such studies should include high-resolution CO maps; infrared observations of cloud complexes; conventional stellar population studies; and x-ray observations of supernovae, close binaries, O stars, and M dwarfs. Improved theoretical estimates of the colors of stellar populations of both very low and very high metal abundances are also vital.

Major advances in the determination of abundances in the interstellar medium will come from UV resonance-line observations from ST and from infrared fine-structure lines; the latter observations require a cryogenically cooled infrared telescope above the atmosphere. The increased light grasp of a large ground-based optical telescope would also be an enormous aid to spectroscopic studies of individual stars in nearby galaxies. Efficient, two-dimensional detectors throughout the infrared from 1- to 40-μm wavelength could play a vital role in mapping the structure of star-forming regions. Millimeter-wave interferometry would be an extremely powerful tool for studying molecular clouds in external galaxies.

13. What are the physical processes involved in the ejection of material from quasi-stellar objects and active nuclei?

A major unsolved problem of quasars is how energy is transferred from the tiny central source to the surrounding expanding radio jets and lobes. These jets and lobes are important both individually and as a measure of the activity of QSO's and other nuclei integrated over time. To understand the origin and evolution of these radio sources, estimates of the relativistic-particle distributions, magnetic-field structure, thermal-particle density, and macroscopic velocity field are required.

VLA maps will yield most of these quantitites under certain assumptions. However, maps with 1-20-arcsec resolution at millimeter wavelengths and below 50 MHz are necessary to set the upper and lower cutoffs in the particle-distribution function. The large millimeter-wave telescopes and interferometers now planned or under construction should be adequate at high frequencies, but ultimately a space interferometer at low frequencies should be considered. X-ray and optical imaging of jets and lobes is needed to study very-high-energy particles and the inverse-Compton scattering of the microwave background by the lower-energy relativistic electrons.

We would also like to understand the relationship between radio jets and lobes and the high-velocity ejected thermal material as seen in broad QSO absorption lines; both of these phenomena seem to require similar amounts of energy for production. Spectra of the absorption lines in the visible with high resolution and signal-to-noise ratio obtained with larger ground-based optical telescopes and in the UV region with ST will be essential here.

14. What are warps and other structures on the outskirts of galaxies telling us about the dynamical history of these systems?

We know already that galactic H I disks, which usually extend well beyond the optical radii, are often curiously distorted or warped. Moreover, the outskirts of galaxies as normal as M81 or the Magellanic Clouds, or as exotic as Centaurus A or Fornax A, display various debris that suggests recent and vigorous interplay with neighbors or fellow group members. Explorations of such faint outlying material promise more than just indirect dynamical probings of any dark halos. Situations like these seem especially valuable as examples of possible accretion or exchange of gas by nearby galaxies, the structural modification of one galaxy by another, and perhaps also the long-term fate of groups of galaxies. Their study should be continued in the 1980's via the most up-to-date 21-cm radio interferometry and deep optical imaging and spectroscopy. The major technical advance here will be the VLA spectral line system.

15. What is the extragalactic distance scale, within 10 percent?

Hubble's constant is currently uncertain by a factor of 2. Improved accuracy is important for determining the physical properties of all extragalactic sources and in determining the age of the Universe. This uncertainty can be resolved in the next decade by utilizing standard candles such as RR Lyrae stars, Cepheids, novae, and so on—all of which can be studied in nearby galaxies with ST.

Three new and completely independent methods for determining distances also remain to be exploited. Two of these are based on supernovae, the third on the microwave diminution in the direction of clusters of galaxies. This last method depends on high-sensitivity, high-angular-resolution x-ray imaging of the intergalactic medium in clusters, plus similar detailed mapping of the microwave diminution at millimeter wavelengths. Improvements in millimeter-wave receivers are required, and ultimately perhaps a large (30-m) millimeter-wave antenna in Earth orbit as well.

III. THE EARLY UNIVERSE, OBSERVATIONAL COSMOLOGY, AND COSMOLOGICAL MODELS

A. Review and Summary

Research in this area during the 1970's was dominated by the impact of the discovery of the microwave background radiation in 1965. Most astronomers take the microwave background as strong evidence that the Universe has expanded from a denser state, for no one has been able to understand how the Universe could, at its present density, have produced blackbody radiation at 3 K. It is just possible that the radiation was produced in stars at a red shift of about 300. If further analysis and tests (of the fine-scale anisotropy) can rule this out, we will know the primeval entropy per baryon and thus a first approximation to the thermal history of the Universe back to the big bang.

During the 1970's, the blackbody nature of the microwave spectrum was generally verified, although small but tantalizing departures from a blackbody curve may have been detected at both long and short wavelengths. Such departures from a pure blackbody spectrum are important for several reasons. They trace the detailed thermal history of the Universe during recombination, are sensitive to inverse-Compton scattering by a diffuse hot intergalactic medium, can reveal emission by stars or warm dust in galaxies at a red shift of about 100, and can reveal any re-ionization of the intergalactic medium at epochs following recombination (see Section VI). For all these reasons it is extremely important to establish the spectral distribution of the microwave radiation with the highest possible accuracy.

The large-scale anisotropy in the background radiation arising from the Earth's peculiar motion was also apparently detected during the 1970's, and the amount was surprisingly large, some 600 km/sec. The limits on the fine-scale anisotropy were set at less than 10^{-4} on an angular scale of 10 arcmin. It is vital to extend these anisotropy measurements to smaller angular scales since this is ultimately the only way to test whether the background consists of point sources. Small-scale irregularities are also our only direct observational link to inhomogeneities in the Universe at decoupling, which in turn are crucial to theories of galaxy formation and clustering. These small-scale irregularities also test for inverse-Compton scattering of the microwave background by hot intergalactic gas in clusters known as the "Sunyaev-Zeldovich effect"; for further discussion see Section VI.

On the theoretical side, the evolution of the mass distribution prior to and through decoupling is currently well understood in the linear perturbation approximation. The challenge now is to assess the possible role of nonlinear effects. There has also been great progress in understanding the expected residue of particle reactions during the high-temperature phase--^{3}He, ^{4}He, d, Li, quarks, baryons--at least within the context of ordinary Friedmann-Lemaitre models. The study of primeval abundances of the light elements is likely to tighten constraints on possible departures from the simplest big-bang scenario.

A coupling between cosmology and elementary-particle physics was one of the most interesting developments of the last decade. On the simplest level, we now realize that neutrino degeneracy, or too many kinds of neutrinos, or neutrino masses in the wrong range of values spoil the modest successes of the simple scenario. Somewhat deeper is the thought that elementary-particle processes may dictate the character of density irregularities present at the big bang. A goal for the 1980's is to learn whether this might have occurred and whether the predictions agree with the demands of the present mass distribution. Many scientists believe that a theory of the big bang itself awaits the unification of quantum and relativity principles.

Considerable progress was made over the past 10 years in extending classical tests of cosmological models to greater lookback times. For example, the red-shift-magnitude relation for galaxies was pushed nearly to a red shift of unity. However, as these observations were being made, it was also gradually realized that these tests are all extremely sensitive to any evolution of the postulated "standard candles." Thus, we have probably learned more about the evolution of galaxies and radio galaxies than about cosmology and cannot yet state definitively whether the Universe is open or closed. We propose a 10-year moratorium on pronouncements as to whether the Universe is open or closed.

Essential to the tests of cosmological models is an estimate of the mean local mass density and a comparison with the density estimates eventually expected from lookback tests. Such local dynamical measures include the careful assessment of galaxy masses, the measurement of peculiar streaming velocities around large concentrations of galaxies, and the application of the cosmic virial theorem. As larger samples of red shifts of nearby galaxies accumulate during the next decade, the power of this approach in modeling the expansion of the Universe will doubtless be exploited extensively (see also Section V).

Three new methods have recently been proposed for obtaining the accurate distance to high-red-shift objects that are necessary for the determination of the deceleration parameter q_0. Two are based on supernovae; one assumes that Type I supernovae are standard candles, the other derives the absolute luminosity of supernovae using the expansion velocity and the effective temperature of the expanding shell. This second approach has a built-in, internal check since the method can be applied to a single supernova at several states during its evolution and, of course, should yield the same distance at all times. The third method is based on the inverse-Compton scattering of the microwave background by hot gas in clusters of galaxies (the so-called Sunyaev-Zeldovich effect). These three methods are promising because they are in principle independent of evolutionary effects, but they all need to be extensively tested on nearby objects at known distances before they can be applied with confidence to distant objects.

The measurement of the local distance scale, or Hubble constant, continued to be a lively source of controversy during the 1970's, and today the true value is probably still uncertain by a factor of 2. However, we look for considerable progress on this problem during the next decade as a result of new or strengthened observational approaches. These include the rotational-velocity/magnitude correlation recently discovered for spiral galaxies, which converts them to standard candles, and the measurement of classical standard candles such as RR Lyrae variables, novae, Cepheids, and H II regions in nearby galaxies with ST. The three new methods mentioned above may also be useful in determining distances to nearby objects during the 1980's.

One might perhaps question the importance of the Hubble constant; after all, it seems easy enough to write all cosmological equations in scaled form to allow for the uncertainty in H_0. However, the value of H_0 strongly affects the assumed luminosities, sizes, and densities of virtually all extragalactic objects and also sets an upper limit to the age of the Universe and clustering time scales for galaxies in the simplest Friedmann models. The assumed ages and spectral distributions for stellar population models in galaxies are strongly affected. The factor-of-2 uncertainty in H_0 also results in an order-of-magnitude uncertainty in the time scales of all density-dependent processes. For the general health of extragalactic astronomy, we think it desirable to know the value of the distance scale to an accuracy of 10 percent. Through the intercomparison of these new observational approaches, this accuracy should be achievable within the next decade.

In the realm of theoretical cosmology, the standard hot big-bang (Friedman-Lemaitre) models have achieved widespread acceptance. These models appear to provide an adequate and simple explanation of the Universe for z = 1 or less (galaxies), z = 1-3 (radio galaxies and quasars), z up to 1000 (large-scale isotropy of the cosmic microwave background), and z up to 10^9 or so (spectrum of the cosmic microwave background, light-element abundances). At even earlier epochs, rival cosmologies may play an important role; these include anisotropic models and baryon-antibaryon symmetric models.

Even on the simplest levels, however, many important theoretical questions remain. What was the origin of the local baryon-antibaryon asymmetry? How was mass distributed (and moving) before decoupling? How did the distribution and motion of matter evolve during decoupling, and how are they related to the distribution of matter and radiation we see today? How do we find ab initio a theory of the big bang itself? All these questions are lively subjects of debate now and are likely to remain so into the next decade.

B. Prospects for the 1980's

1. The Spectrum and Isotropy of the Microwave Background Radiation

Because the microwave background provides by far our most important observational probe of the physics of the early Universe, it is vital to improve the accuracy of the measurement of the spectrum and to place more stringent limits on the large- and fine-scale anisotropy. Deviations from a blackbody spectrum are interesting at both short and long wavelengths. Deviations at wavelengths shortward of the spectral peak could signal dust emission by primeval galaxies or inverse-Compton scattering of the microwave background by a re-ionized intergalactic medium after decoupling. At long wavelengths, excess flux could come directly from thermal bremsstrahlung by such a re-ionized intergalactic medium. It could also indicate injection of energy into the plasma just before decoupling, for example by matter-antimatter annihilation or viscous damping of turbulence.

The Cosmic Background Explorer (COBE) satellite will ultimately provide excellent measurements of the spectrum from 0.3 to 5 mm and of the large-scale anisotropy (on a scale of 10° or greater) from 3 to 1.5 cm, though interim measurements by balloon are also to be encouraged. Virtually all of the existing spectral measurements at wavelengths longward of 1 cm are quite old and could be substantially improved today; beyond about 10 cm, the atmosphere poses little problem, and existing ground-based telescopes could be used. Substantial improvement is possible from the ground in the 3-10 cm range as well.

Better limits on the fine-scale anisotropy are probably the single most important measurement still to be made on the microwave background. These observations are needed to establish beyond doubt that the background is not composed of point sources. They may also detect inhomogeneities at decoupling, which foreshadowed the eventual formation of galaxies and clusters of galaxies. The measurements should be made at all practicable wavelengths and on angular scales down to at least 1 arcmin, the approximate size of a protocluster of galaxies at $z = 100$. Angular resolution on this scale is also necessary to exploit the Sunyaev-Zeldovich distance measurement to nearby clusters (see Section VI).

Measurements of this type are currently limited almost equally by receiver noise, atmospheric effects, and variable ground pickup in the sidelobes of the telescope. Substantial improvements in receiver sensitivity and sidelobe noise are possible from the ground and should be encouraged. However, the atmosphere may well prove to be the limiting factor for ground-based work; if so, an ultimate exploration of the micro-wave background might utilize a large radio antenna in Earth orbit having an aperture of about 30 m. With a surface usable down to wavelengths of 1 cm or less, it would have an angular resolution of 1 arcmin, small enough to detect inhomogenei-ties on the scale of protoclusters.

2. Abundances of the Light Elements

The deuterium-to-hydrogen (D/H) ratio places limits on the mean mass-density of the Universe, since deuterium production in the big bang was very sensitive to density, and no stellar or supernova environment has been identified in which deuterium can be formed after the big bang in significant amounts. However, deuterium is generally destroyed in passing through stars, so unprocessed primordial material is required for measurement. A good place to look for the primordial ratio may be in those gas clouds producing the large number of Lyman-alpha absorption lines in quasars. The D/H ratio involves a splitting of about 80 km/sec and, for a few quasars, could be measured either from the ground or in space with spectral resolutions of about 6000 or greater. Increased aperture, either in space or on the ground, would help greatly.

In principle, the ^{3}He/^{4}He ratio also places an inter-esting constraint on big-bang models; however, the interpreta-tion is complicated by possible nucleosynthesis in stars. The splitting will also be about six times smaller. Since the He II line at 304 Å is not accessible from the ground in any known QSO, only an instrument operating in space with a spec-tral resolution of about 10 km/sec can make this observation. (Such an instrument is currently being built for ST, but a substantial increase in sensitivity is necessary.)

3. The Diffuse Gamma-Ray Background

Grand Unified Theories with spontaneous symmetry-breaking in the very early big bang can lead naturally to a baryon-symmetric cosmology with a domain structure. The symmetry is broken randomly in independent domains favoring neither a baryon nor an antibaryon excess on a universal scale. As a result of matter-antimatter annihilation at the domain boundaries, this theory predicts a smooth, diffuse gamma radiation with a unique energy spectrum. There are some indications that this radiation may exist, but gamma-ray observations in the energy range near 0.5 MeV are necesary to verify this possibility.

IV.   THE DISTRIBUTION OF MASS ON SCALES GREATER THAN 10 kpc

A.   Review and Summary

Since the large-scale clustering of matter may be a more recent event than is the clumping of matter into galaxies, stars, and so on, it may be easier to trace the large-scale clustering back to its origin and discern what it implies about the initial structure of the Universe. Furthermore, the dynamics of galaxies viewed as mass points provides a unique opportunity to weigh the Universe locally and thus obtain a measure of the total mean mass density that can be compared with the classical lookback determinations of q_0 (see Section III).

Research in this area during the last decade has emphasized a wholly new approach: statistical analyses of the galaxy ensemble that bypasses the need to label individual galaxies as members of particular groups and clusters. We now have fairly well-defined n-point correlation functions, luminosity functions, and multiplicity functions for galaxies and groups of galaxies. As a result, a picture of the space distribution of galaxies is emerging. The sizes of the great clusters and the extent of the "holes" between clusters are larger than many would have imagined in 1970.

The analyses up to now have used mainly galaxy angular positions and apparent magnitudes. An obvious next step is to add red shifts. This would greatly enhance the visibility of the clustering by reducing the confusion of galaxies seen together in accidental projection and is also indispensable to the study of cluster dynamics. During recent years, a start has been made on a large-scale red-shift survey for the brighter galaxies, but it is discouraging that some of the catalog red shifts are still uncertain by as much as 300 km/sec. Errors of this magnitude render velocities almost useless for the dynamical study of the peculiar motions of galaxies relative to the Hubble flow.

Several promising starts were made in applying statistical analyses to the joint distributions of positions and red shifts. The situation is still unstable because the red-shift data base remains small. Nevertheless, these treatments confirm the results of more conventional analyses on individual binary galaxies, groups, and clusters (all of which were also further refined during the 1970's) and suggest that as much as 90 percent of the Universe consists of essentially invisible, nonluminous matter quite different in its properties from the gas and stars that comprise the visible portions of galaxies.

On the largest scales, the distribution of extragalactic radio sources was found to be remarkably uniform (isotropic, random Poisson), and this fact, together with the isotropy of the diffuse x-ray background, is the strongest direct evidence supporting the conventional assumption that the Universe is very nearly homogeneous on the scale of the Hubble length, c/H_0. The possible large-scale clumping of quasi-stellar objects (QSO's) remains controversial because of variable efficiency of detection from field to field.

Some theoretical work has been done on how galaxy clustering might have evolved in an expanding world model. Most workers in this field agree that the gravitational-instability picture yields a reasonable first approximation to the observations, but probably most would also agree that the subject is still in a primitive state awaiting a firmer observational and theoretical foundation.

Virtually all of the work on galaxy distributions to date has assumed that the galaxies are bona fide tracers of the total mass distribution. However, there are disturbing indications from the study of individual binaries, groups, and clusters that the proportion of matter in the nonluminous component varies widely from aggregate to aggregate by as much as a factor of 100. The highest mass-to-light ratios appear to occur in the largest clusters, where invisible matter approaches 95 percent of the total mass.

The observations raise the issue of the distribution of the nonluminous component relative to luminous matter, and the flat rotation curves discovered around galaxies during the past decade have assumed the highest importance in this connection. They suggest that the total mass distributions of ordinary spirals extend far beyond the optically visible interior regions. In short, the rotational data strongly imply that at least a sizable fraction of the nonluminous matter in the Universe resides in the outer regions of individual galaxies themselves. Optical searches for this material have been inconclusive, confirming only that it must have a mass-to-light ratio in excess of several hundred times that of the Sun.

Together with the discovery of flat rotation curves came the realization that outer H I disks in spiral galaxies are often significantly warped. If all the mass of the galaxy were distributed in a disk, the resultant large quadrupole moment would cause these warps to precess differentially and hence to smear out and disappear in an embarrassingly short time. Simply the existence of these warps suggests that the bulk of the mass of the galaxy may have to reside in a spherical or even triaxial halo. Since such halos are not visible optically, the material must have high mass-to-light ratio and is therefore likely to be related to, if not identical with, the nonluminous material binding groups and clusters. The warp problem may thus be intimately intertwined with the study of the nature and distribution of the mass of the Universe in general.

In summary, this past decade has seen a virtual revolution in our analysis of the large-scale mass distribution and has strengthened evidence in favor of a dominant, nonluminous component of matter in the Universe. The major questions facing us now concern the origin of the inhomogeneities that gave rise to groups and clusters of galaxies, the influence of the unseen mass component of the dynamics of galaxy formation and clustering, and the most difficult question of all: What is the nature of this ubiquitous but invisible stuff?

B. Prospects for the 1980's

1. Red-Shift Surveys of Nearby Galaxies

To study the local distribution of matter in the Hubble flow, a sample of nearby galaxies complete to apparent blue magnitude 15 is required, preferably with errors of 75 km/sec or less. Measurements such as these require generous allocations of time on moderate-sized telescopes, plus high-efficiency multiplexing spectroscopic detectors.

2. Theoretical Studies of Galaxy Clustering

To analyze patterns of galaxy peculiar motions, we need realistic models of galaxy clustering that explicitly include any nonluminous component. A large N-body computer could make an important contribution to these theoretical studies.

3. Deep Surveys of Galaxy Properties

Deep surveys of galaxy colors, positions, and red shifts are vital for study of the degree of isotropy in the distribution of distant galaxies, the evolution of galaxy clustering, and the evolution of galaxies themselves out to red shifts of order unity. To reduce the ambiguity in the interpretation of the number-magnitude counts, we need broadband colors of each galaxy counted, together with fair samples of red shifts at each of several magnitudes down to $m_v = 24$. With present equipment and techniques, measurements of such faint objects are currently too slow to make large-scale surveys practical. The development of multiplexing red-shift detectors capable of measuring many galaxies simultaneously in the same field is crucial to the success of these surveys. A larger ground-based optical telescope would be important here as well.

4. Study of Individual Systems of Galaxies

Study of individual structures in the galaxy distribution will continue to be worthwhile. We noted above that the cores of

great clusters have exceptionally high mass-to-light ratios,
whereas smaller groups like those in the outskirts of our own
Local Supercluster appear to have a lower value of this ratio.
The great question now is: How does the mass-to-light ratio
vary with distance from the centers of great clusters? Finding
an answer depends on accumulating statistics from a fair sample
of cluster centers of the galaxy velocity dispersion and mean
streaming velocity as a function of distance.

As a second example, the filaments of galaxies seen in the
Lick galaxy counts by Shane and Wirtanen--and in shallower
samples, too--should be studied in detail to distinguish true
physical structures from statistical accidents. If real, the
filaments may reveal much about the shapes of inhomogeneities
in the early Universe.

Each of these projects clearly demonstrates the need for
more and better red-shift measurements of nearby and distant
samples of galaxies.

5. Deep Surveys for Quasi-stellar Objects, Seyfert Galaxies, X-Ray Sources, and Radio Galaxies

These surveys are likely to remain our only probe of clustering
in the Universe in the red-shift range z = 1.5-4. We need to
know whether the clustering of QSO's is similar to or greater
than the very weak clumping among distant radio galaxies. Deep
radio surveys may prove to be the best probes of the mass
distribution on scales greater than 100 Mpc because interfer-
ence by our Galaxy is minimal and the sources are widely spaced
markers.

Potentially important techniques in this area include deep
x-ray and radio surveys at high spatial resolution and optical
objective-prism surveys for QSO's on the ground and from space
(see Section VII for further discussion of QSO surveys).

6. Small-Scale Anisotropy in the Microwave Background

As discussed in Section III, small-scale irregularities in the
microwave background are potentially important tracers of
inhomogeneities in the mass distribution and motions at high
red shifts. Observations of the fine-scale structure in the
background are thus our best hope of finding the inhomogeneities that gave rise to the clusters of galaxies that we see
today. The techniques discussed in Section III for the
measurement of small-scale inhomogeneities in the 3 K
background apply equally here.

7. Warps and Halos of Galaxies

Understanding the outer rotation curves, warps, and halos of
galaxies has extremely high priority for the next decade. The

principal stumbling block in this area now is theoretical ignorance of these structures, but theorists could profit from additional 21-cm observations of galaxies that cannot at present be reached with the Westerbork array because of their small angular size or southerly declination. With its higher resolution and good antenna pattern even in the south, the VLA will represent a huge step forward in 21-cm work once its planned spectral-line system is completed. Since the geometry of collisions and mergers of galaxies may be sensitive to the gravitational influence of unseen mass in a halo component, theoretical models of collisions incorporating halos should soon be compared with high-sensitivity H I observations of colliding galaxies.

8. Dynamics of Globular-Cluster Systems

Globular-cluster populations offer an ideal probe of the mass distribution in the outer regions of nearby galaxies. They can be followed to very large radial distances, far beyond the point where the surface brightness of the underlying spheroidal component becomes too faint to study. Tidal radii of globular clusters can be measured with Space Telescope. Radial velocities will give a completely independent datum needed to determine orbital eccentricities. However, with visual magnitudes generally in the range m_v = 20 or fainter, globular clusters are too faint for current red-shift techniques. An areal red-shift multiplexer and a larger ground-based optical telescope are crucially important here.

V. THE FORMATION AND EVOLUTION OF GALAXIES

A. Review and Summary

Many years from now, the 1970's are likely to be remembered as a period when the full complexity of galaxy evolution was glimpsed for the first time. On the observational side, new detector technology produced an extraordinarily rich banquet of observational data--in some cases bewilderingly complex, in other cases reassuringly regular. Armed with computer simulations of a sophistication never before seen in galactic studies, theorists countered with an imposing array of imaginative theories, in many cases based on physical mechanisms rarely applied to galaxies before. As a result of this activity, there exists a widespread feeling that we are now in the midst of a major revolution in our thinking about galaxies; as of this writing, however, very few of the basic questions have yet been settled. During the 1980's, we expect a continued high level of activity in this field, with the real possibility that by the end of the next decade many of the major pieces of the puzzle will have fallen into place.

1. Theoretical Underpinnings

Unfortunately, we still do not have any _ab initio_ theory for the formation of protogalactic structures because we simply do not know what the Universe was like before there were galaxies. Measurements of spatial variations in the microwave background are of vital importance here since these small-scale fluctuations may well prove to be our only direct observational link to galaxy formation.

Most workers have circumvented this obstacle to our knowledge by postulating various possible protogalactic structures and working out the resultant processes of collapse and coalescence using impressively detailed computer simulations. However, the enormous complexity of three-dimensional gaseous hydrodynamics has not yet been modeled realistically. The accumulating observational evidence seems to favor protogalactic structures that are largely gaseous as opposed to stellar, but this point is still controversial. The more fundamental question of how the presence of the unseen material might have affected the initial formation process has hardly been considered. The crucial issue of time scales is also still very uncertain: At what epoch did the coalescence into galaxies begin, and how rapidly did it occur? What coalesced first--the ordinary gas or the massive, unseen stuff? What came first--star clusters, galaxies, or clusters of galaxies?

2. Heredity versus Environment

The major revolution in our thinking about galaxies during the last decade stemmed from the new realization that, after their formation, galaxies are not necessarily the "island universes" they seemed to be, isolated from one another and from the rest of the Universe. On the contrary, they appear to interact with one another and with their environment in a number of ways that may profoundly affect their evolution.

The first breakthrough in this area was the demonstration that many galaxies, even those of the most bizarre appearance, could be understood simply as the result of the two galaxies in violent collision, tidally distorting one another. Quickly after this discovery came the finding that close encounters could lead to outright mergers and that, statistically speaking, this might have happened rather frequently since galaxies were first formed. Mergers may also be enhanced by the process of dynamical friction against the background mass in clusters of galaxies, which should lead eventually to the formation of a giant cD galaxy at the cluster core. If there is any problem with this picture it is the fact that dynamical friction seems to be too efficient. For example, it is hard to understand theoretically how binary galaxies or compact groups could persist for any length of time in orbits with mutually penetrating massive envelopes, yet real galaxies seem to manage it quite nicely. We urgently need to know whether our theoretical time scales for dynamical friction are reasonable.

Several mechanisms were suggested that could influence the gas content of galaxies, a quantity closely correlated with Hubble type. These processes include sweeping by galactic winds, ram-pressure stripping and evaporation by a hot inter-galactic medium (IGM), heating and stirring by supernovae explosions, and recent infall of intergalactic gas or gas-rich dwarf galaxies. Moderately convincing observational evidence was claimed for all of these possibilities. Attempts to discriminate between them will continue, primarily utilizing x-ray observations of the hot IGM and 21-cm data for neutral hydrogen in and around galaxies.

The question of heredity versus environment as the control-ling factor in galaxy evolution was crystallized by studies of the great clusters of galaxies. Much evidence was uncovered that bolstered the earlier suspicion that these clusters repre-sent an evolutionary sequence, from those just now collapsing to those already well into dynamical equilibrium. The crucial fact here for galaxy evolution is that morphological types of galaxies in clusters also vary systematically along this se-quence, from spiral-rich clusters at the one end to clusters dominated by ellipticals and SO's at the other. It has been tempting to attribute this progression of Hubble types to the stripping of spirals by the hot IGM, producing SO's that in turn perhaps collide or merge to form ellipticals. However, other evidence based on comparisons of the underlying mass distributions between spirals and SO's and on more precise studies of the distribution of morphological types within clusters already seems to require significant modifications to this simple picture. This particular question—the influence of environment on Hubble type—is likely to remain one of the most active areas of research well into the 1980's.

3. Lookback Studies of Galaxies

Lookback studies of galaxies to significant red shifts have only just begun during the last few years, but they have already uncovered surprising results. In studies of distant clusters and the general field at faint magnitudes, an unexpectedly large population of galaxies was found with mean colors much bluer than those of nearby galaxies. It seems likely that some strong evolution in galaxy properties is being detected, possibly signifying the rapid transmutation of spirals into SO's in at least the dense clusters studied. Whatever the true explanation, these results suggest that lookback studies of the stellar populations, morphological types, and magnitudes of galaxies will gradually elucidate the factors governing galaxy evolution, especially when pursued in conjunction with studies of the evolution of galaxy clustering. The high spatial resolution of Space Telescope (ST) offers special promise here in studying the morphology of extremely distant galaxies.

4. Elemental Abundances and Stellar Populations

Composition gradients were discovered or confirmed in the inner regions of virtually all galaxies studied, in rough agreement with the predictions of gaseous-collapse models. Probable metallicity variations among elliptical galaxies were found to correlate with luminosity and axial ratio in a complex way. These, too, seem easily explained only if protogalaxies were largely gaseous.

Modeling of stellar populations along the Hubble sequence generally confirmed the earlier view that the sequence is basically an ordering according to the rate of recent star formation. It is becoming more and more obvious that if we are ever to understand Hubble types, we must first discover what factors controlled the gas content and the star-formation rate in both the recent and the distant past.

Recent x-ray observations have revealed interesting facts about stellar populations in external galaxies. For example, x-ray emission from individual supernova remnants was detected; statistics concerning these objects should help to fix the supernova rate more accurately, determine the masses of supernovae progenitors, and study the interaction between supernovae and the interstellar medium. M dwarfs have also been found to be strong x-ray emitters; x-ray observations of nearby galaxies may thus be able to show whether the unseen matter surrounding galaxies is composed of low-luminosity M-dwarfs. A strong class of x-ray sources has been discovered in the inner bulge of M31; these sources are presumably stellar or have stellar progenitors, but their nature remains a mystery. Finally, the most luminous mass-exchange binaries in the Large Magellanic Cloud appear to be somewhat brighter than their galactic counterparts, perhaps because the upper limit to their luminosity set by radiation pressure is higher because of their lower metal abundance. If this effect could be allowed for, x-ray binaries might become yet another new standard candle in the 1980's.

5. Dynamical Properties

Advances in galaxy dynamics during the last 10 years made it clear that considerable richness and novelty may still be found in the traditional celestial mechanics that deals primarily with gravitation. This truism spans the whole extragalactic domain, from great clusters of galaxies to presumed black holes in galactic nuclei (see Section VII). Even for allegedly normal galaxies, however, an account of the progress since 1970 on such topics as density waves, bar mechanics, global instabilities, and tidal interplay sounds somewhat like an advertisement for the pleasures of space travel under the influence of gravity.

We shall cite the theoretical accomplishments first, beginning with spiral structure. From several decisive 21-cm studies, it was established firmly during the past decade that at least the major spiral arms of galaxies like M51 and M81 are wavelike in character. It was also shown by theorists that shearing gas disks in galaxies are very prone to develop spiral shock waves when forced by quite modest but rotating disturbance potentials, especially when bars were imposed. On the other hand, whether or not these structures are truly long-lived or merely transient responses is still uncertain.

The main triumph of N-body simulations of rotating galaxies was the development of massive oval or barred structures in the interiors of the model galaxies. In all probability, these long-lived computed shapes capture the essence of the bars or ovals actually observed in the interiors of real SB or SAB galaxies. At the same time, these calculations disclosed very serious difficulties in plausibly stabilizing "cool" disks against fierce spiral instabilities. Possible solutions put forward so far have emphasized the presence of a massive, spherical, dynamically inert halo of low luminosity--unseen matter again. Much work and testing remains before theorists can feel secure about how galaxies stabilize themselves, with or without halos.

Three shocking observational results emerged during the 1970's: the flat outer rotation curves, the common occurrence of warps, and the slow rotation of elliptical galaxies. The first two have been mentioned earlier; the third doomed most existing models of ellipticals and even cast doubt on their basic shapes--whether prolate, oblate, or triaxial.

6. The Interstellar Medium

We have already mentioned some of the new ideas about mechanisms that might control the interstellar medium within galaxies. Other recent observational results include several high-resolution 21-cm maps of H I distributions, which showed that H I often extends to several times the optical radius of a galaxy. Since ablation by an intracluster medium ought to be more effective at large radii, where the gravitational restoring force is weak, 21-cm maps of clusters with the VLA may reveal individual galaxies with abnormally small H I extent. If so, we will perhaps have found galaxies currently being shorn of their gas content.

Carbon monoxide and other molecules were discovered for the first time in external galaxies during the 1970's. In the

Milky Way, CO studies have played a major role in localizing sites of active star formation. CO maps of galaxies over a wide variety of Hubble types will illustrate how star formation is triggered in other types of galaxies.

B. Prospects for the 1980's

1. Lookback Studies of Galaxy Evolution

These observations should include galaxy surveys to the deepest possible magnitude to measure luminosities, colors, counts, and positions; red shifts of a fair sample at each magnitude and color are also needed to determine the distance scale. Such surveys should be carried out in conjunction with the clustering studies outlined earlier to look for changes in galaxies during the clustering process. Morphological studies with ST of galaxies at z = 0.5 or greater have high priority.

The ultimate limit in red shift for these studies lies in the range z = 1-2, somewhat higher if the luminosity of galaxies was much greater in the past. For measurements of integrated magnitudes and colors at wavelengths beyond about 0.9 μm, a large (10-m or greater) ground-based optical telescope and ST would be comparable. This equality is due to the fact that the greater aperture of the ground-based telescope is offset by the brighter night-sky airglow bands, which are absent in space. For wavelengths between 0.3 and 0.7 μm, the ground-based telescope is substantially more powerful than ST. The higher angular resolution of ST makes it preferable for morphological studies, however, while the higher photon flux of the ground-based telescope is indispensable for accurate red shifts. The development of red-shift multiplexers and highly efficient two-dimensional detectors in the range 0.3-2 μm is also essential to all aspects of this program, which strains the limits of every facility.

Whether or not these deep surveys will actually detect primeval galaxies cannot be predicted, since no one has a very good idea what these objects should look like or what their luminosities will be. Deep exposures with ST will certainly be analyzed for likely candidates. Since some models of primeval galaxies predict a large initial burst of x rays from early generations of supernovae, deep x-ray surveys with high angular resolution ought not to be overlooked; the optimal energy band for such a search is near 1 keV. Stellar radiation from an early, intense period of star formation may be thermally reradiated at infrared wavelengths by enveloping dust clouds; advances in infrared sensitivity beyond 10 μm will play a crucial role in searches at these wavelengths.

2. Star Formation and the Interstellar Medium

a. Theory

Our continuing lack of a theory for star formation is a severe handicap to models of galaxy evolution. We need expressions for the rate of star formation and the initial mass function as functions of the local gas parameters. This is a tall order, and one would be surprised if final answers were forthcoming over the next 10 years. More realistically, one might hope for substantial progress on the more straightforward questions of mass loss among massive stars, the evolution of stars of low metal content, and on the nucleosynthetic yields of helium and the heavy elements.

b. CO Maps of Galaxies

High-resolution CO maps of other galaxies will allow us to study the distributions of giant molecular cloud complexes with respect to spiral arms, H II regions, dust lanes, and supernova remnants. The 25-m millimeter-wave dishes now proposed in the United States and Germany will have angular resolutions of about 20 arcsec, adequate for studies of Local Group galaxies. For more distant objects, millimeter-wave interferometers and low-noise receivers are required.

c. Infrared Continuum Fluxes from Cloud Complexes

Infrared continuum measurements are needed to determine the integrated emission from early-type stars, including those obscured by dust; these data will in turn help fix the star-formation rate and the initial mass function. Observations between 10 and 100 μm are necessary, and they can be made at the required level of sensitivity only with a cryogenically cooled infrared telescope in space. Two-dimensional panoramic detectors for direct imaging over this wavelength range would speed measurements enormously.

d. Extragalactic Supernova Remnants

Extragalactic supernovae have several potential applications. X-ray observations of many supernova remnants in the Large Magellanic Cloud have shown, for example, that remnant diameters are inversely correlated with the density of the surrounding interstellar medium. This suggests a sensitive means of probing the density of the interstellar medium in nearby galaxies. Supernova statistics also tell us much about the initial mass function for new stars. Furthermore, the interiors of all old remnants constitute the hot-gas phase in the interstellar medium, which has an estimated temperature of

10^6 K. The fraction of the volume occupied by this phase in our own galaxy is highly controversial. Mapping external galaxies at high spatial and moderate spectral resolution should determine whether there is a hot phase and how its temperature and density vary with position in the disk; this will require a spectral resolution in x rays of 5 or greater over a wavelength range of 0.1-2 keV.

e. Infrared Spectroscopy of the Interstellar Medium in Galaxies

Infrared spectroscopy of vibrational and higher rotational lines of H_2 (2 and 12 µm), CO (2 and 5 µm), OH and H_2O (50-100 µm), all of which are excited by shock waves, would provide diagnostics of energy injection into molecular cloud complexes by stellar winds, H II regions, and supernovae. The temperature of the cold, neutral medium whose density is found from 21-cm maps can be inferred from fine-structure lines of C II (156 µm) and Si II (35 µm), and the arm and interarm properties of the cold gas can be compared. Any ionized medium near 10^4 K will emit Ne II at 12.8 µm, N II at 122 µm, and other near-IR fine-structure lines; these lines will be especially useful in determining gas temperatures because they are virtually insensitive to density. Furthermore, ionization species such as S III can be observed, thus removing the need for model-dependent corrections to abundances for unseen ionization states.

As this brief summary illustrates, the infrared region offers many exciting possibilities for spectroscopic studies of the interstellar medium that have hardly yet been exploited. In all of these applications, however, high spatial and high spectral resolution are required. Some of this work can be carried out in spectral windows from the ground or from high-flying aircraft, but a cryogenically cooled infrared telescope in space would provide an enormous gain in sensitivity. Two-dimensional panoramic detectors throughout this spectral region would also speed measurements greatly.

f. Ultraviolet Absorption-Line Spectroscopy of Galaxies

These studies using ST could do for external galaxies what the Copernicus satellite did for the solar neighborhood of our Galaxy. With quasars as background sources, abundances in gaseous coronas around galaxies at very large galactocentric distances can be measured. The relevant spectral features are resonance lines of several abundant atomic species located in the UV region; a new UV echelle spectrograph for ST would be ideal for this applications.

One especially significant region is the spectral range 912-1100 Å, where very important transitions--including those of molecular hydrogen, deuterium, O VI, and C III--are located. This region will not be accessible with ST, and new instrumentation designed for Shuttle payloads is required.

g. H I Maps of Galaxies

There is a lot of life left yet in this old warhorse, which will gallop afresh when the spectral-line system for the VLA begins full operation. With a spatial resolution comparable with that of optical telescopes and an excellent beam pattern over three quarters of the entire sky, the VLA will produce H I maps of unprecedented quality--the ultimate data for comparison with theories of spiral structure. The increased sensitivity of the VLA will allow measurements of rotation curves and warps to greater radii, the H I distribution and kinematics in cases of suspected accretion of intergalactic H I clouds, and perturbed H I distributions in interacting galaxies. H I maps of individual galaxies in clusters of varying stages of dynamical evolution should tell us directly how galaxies in clusters came to be gas-poor. The VLA spectral-line system is clearly an extremely important tool for extragalactic astronomy during the 1980's.

3. Theoretical Studies of Galactic Dynamics

Two areas here seem certain to show rapid progress during the next 10 years. The first is understanding the global instabilities of disks, which would obviously yield many side benefits. We will know then, for instance, how much we really need those invisible halos for stability, what it takes to make bars, and why spiral arms do not sprout all over the place. Computed gravitational interactions of stable disk models should also prove one of our best tools for probing the extent and shapes of the halos that we cannot see and for checking whether mergers can actually produce the slow rotations recently discovered in elliptical galaxies.
This leads us to the second major area: understanding the halos and warps. It is clearly the other important direction in which galactic dynamics will be tugged in the coming years. Here are a few typical questions: What sort of "isothermal" statistical mechanics of those unseen halos invites the flat rotation curves? Could the warps result from the visible galaxies remaining tilted with respect to somewhat flattened halos? Might it not be possible for two galaxies to orbit one another steadily within a common halo and yet avoid dynamical friction?

4. Abundance Studies

Future work in this area will depend primarily on long-slit spectroscopy of galaxies, for which the availability of efficient two-dimensional detectors in the region 0.3-2 μm is vital. The study of metallicities, star by star and cluster by cluster, in our own Galaxy has inspired fairly detailed scenarios for the early collapse and nucleosynthetic enrichment of the Milky Way. The recent discovery of variations in metallicity from star to star within globular clusters has added an interesting new dimension to this problem.

Similar measurements based on broadband colors of individual stars and globular clusters need to be made in other galaxies using ST. An areal red-shift multiplexer coupled to a larger ground-based optical telescope would allow detailed spectroscopic studies of the nearer globular cluster systems, enabling abundances of individual elements to be determined. Perhaps the missing stars of essentially zero metallicity will be found at last among globular clusters at extremely large galactocentric distances in other galaxies.

5. The Ages of Galaxies

Debate continues on the ages of galaxies, from ellipticals and dwarf spheroidals on the one hand to the disks of spirals and the Large Magellanic Cloud on the other. ST will provide new information in the form of accurate H-R diagrams below the main-sequence turnoff for the Magellanic Clouds and nearby dwarf spheroidal galaxies; these should provide age estimates based on cluster-dating techniques. Lookback studies of the color evolution of distant galaxies using ST and large ground-based optical telescopes will place important constraints on the epoch of major star formation in normal galaxies as a function of Hubble type.

VI. THE INTERGALACTIC MEDIUM

A. Review and Summary

During the 1970's a hot intergalactic medium (IGM) was detected in groups and clusters of galaxies. Whether a diffuse gas of lower density also exists in space between groups and clusters is still an open question, although meaningful limits have been placed on the allowed temperature and density of such a phase. We first consider results on the cluster IGM and second the question of a more widely distributed IGM.

During the last 10 years, extended x-ray emission was discovered from clusters of galaxies, thereby confirming the presence of a hot, diffuse gas that had been suspected from the shapes of head-tail radio sources. Thermal bremsstrahlung seems to be the primary x-ray emission mechanism. The mass of gas in the cluster cores is comparable with the luminous mass in galaxies and is therefore substantial, although not sufficient to bind the clusters. Iron emission lines are present, from which the iron abundance is deduced to have approximately the solar value. This finding strongly suggests that the gas is not primeval but has experienced nucleosynthetic processing in stars.

X-ray observations of small groups of galaxies showed that they, too, are sometimes luminous, extended x-ray sources. Since the cooling time for the gas in such groups appears to be quite short, we may be viewing these groups at a late stage in their evolution.

The x-ray morphology of galaxy clusters is correlated with their dynamical age, as inferred from the distribution of galaxies within the cluster. In irregular clusters, which are probably still in the initial stages of collapse, the x-ray emission is dominated by gas clumped around individual galaxies. In regular clusters, which are in presumed dynamical equilibrium, the x-ray emission appears to come from a smoothly distributed intergalactic medium. Apparently the x-ray morphology thus offers the means for classifying clusters, mapping their gravitational potential, and examining the state of their dynamical evolution. Since cluster x-ray emission has already been detected out to a red shift of z = 0.75, we should be able to follow the course of this evolution over a large span of lookback times.

There is the possibility that diffuse x-ray emission was detected from superclusters of galaxies, although the point remains uncertain owing to the difficulties of measuring large, diffuse sources. These observations urgently need to be confirmed. If real, the detection is interesting because it suggests the opportunity for studying the gas distribution in the halos of clusters as well as in the space between clusters. The detected mass is dynamically significant and may be sufficient to bind the superclusters, although not to close the Universe.

Radio galaxies in sparse clusters appear to require external pressures similar to those in rich clusters, implying that at least pockets of fairly dense, hot gas exist in those smaller aggregates, too. This idea is consistent with the high x-ray luminosities observed in some small groups of galaxies. More surprisingly, very large isolated radio galaxies with diameters of 1 Mpc or greater have been discovered even in regions of little or no clustering. If thermally confined by a gas with a temperature of 10^8 K or less, these sources would need to be surrounded by a medium with a density of at least 10^{-29} g/cm^3. If such a gas fills all space uniformly, this medium is close to the closure density of the Universe.

We turn now to the question of a uniform diffuse medium outside of clusters. Such a uniform medium is potentially important for several reasons. It could contain a significant fraction of all the mass in the Universe, and it also acts as a "cosmic calorimeter" for all the energetic phenomena in the Universe. It may also have been important during the formation of galaxies.

Observational evidence regarding such a medium is still fragmentary. During the past decade, background radiation at hard x-ray wavelengths was detected, which was found to be quite uniform and isotropic on large angular scales. Its spectrum from 3-50 keV is consistent with emission from a uniform IGM at a temperature of 5×10^8 K. Note, however, that recent high-resolution High Energy Astronomical Observatory-2 (HEAO-2) observations suggest that a large fraction, and perhaps all, of this background is produced by quasars (see Section VII).

An ionized IGM might be detected through its distortion of the cosmic microwave background radiation, as bremsstrahlung emission could increase the observed flux above the blackbody curve at wavelengths below 20 cm. Moreover, if the IGM is very hot, inverse-Compton scattering could decrease the observed flux at wavelengths shorter than 1 mm. Measurements at these short wavelengths during the last decade established that the observed flux differs from the expected Planck curve by less than 10 percent. For an assumed IGM temperature of 5×10^8 K (the temperature derived from the hard x-ray background), the allowed mass in the IGM is less than 0.4 times the critical density of the Universe.

Rocket and satellite observations of the far-ultraviolet background (1000-3000 Å) placed strong limits on emission by the He II line at 304 Å, thus eliminating models in which the IGM has a temperature of about 80,000 K and a density equal to the critical density.

Searches for a neutral medium in the form of H I clouds have been negative except near interacting galaxies. In principle, a neutral medium can also be detected by Lyman-alpha absorption in the spectra of distant objects. This test was extended to quasars with a red shift of 3.5. No evidence for a smoothly distributed medium was found; thus, any diffusely distributed IGM must be ionized out to this red shift. However, discrete Lyman-alpha absorption lines have been known in quasar spectra for some time, and recent tentative evidence suggests that at least some of these are due to intervening H I clouds at high z. Whether these clouds are truly intergalactic or are merely associated with distant galaxies seen along the line of sight is not yet certain.

B. Prospects for the 1980's

1. Cluster Evolution

Several important questions concerning the evolution of the IGM in clusters of galaxies seem ripe for pursuit during the next

decade. These questions are all intertwined with the evolution
of clustering in general and with environmental influences in
clusters that may modify Hubble types and otherwise affect the
evolution of individual galaxies. The cluster IGM thus becomes
an important tool for studying the more general questions of
gravitational clustering and galaxy evolution.

X-ray observations are especially powerful in this connection because of their great range in red shift. The next-
generation x-ray telescopes already being planned should be
able to detect clusters out to a red shift of 4 and iron-line
emission to a red shift of 2. Thus the actual epoch of cluster
formation might be reached in deep x-ray surveys, particularly
since known x-ray clusters show two distinct signatures:
extended size and iron-line emission. Observations of the
x-ray luminosity, angular distribution, and iron-line strength
can tell us how the cluster potential, gas content, and iron
abundance evolved with time. Models for the stripping of
spirals to make S0's can be tested, and the role of cD galaxies
in cluster evolution can also be better determined.

It will be particularly important to compare the x-ray
properties of large clusters with those of small groups, where
environmental influences on galaxies are not expected to be so
great. At the same time, we may learn the means by which these
smaller aggregates generate large x-ray luminosities from the
high-temperature gas.

Much of this work was begun by the Einstein x-ray observatory (HEAO-2), but higher spatial resolution and higher
sensitivity will be necessary to reach very distant clusters
and to obtain high spatial resolution on local clusters.
Observations from 10-100 keV are also necessary to detect
possible nonthermal contributions to the cluster x-ray
luminosity, such as inverse-Compton scattering of the microwave
background, which is predicted to be a larger fraction of the
cluster x-ray flux at high red shifts.

2. The Sunyaev-Zeldovich (S-Z) Effect

This effect is the diminution of the microwave background by
inverse-Compton scattering by the hot cluster IGM. The
fractional intensity change for microwaves is approximately
proportional to the integral of $n_e T$ along a line through the
cluster, where n_e is the electron density and T is the gas
temperature within the cluster. By contrast, the x-ray emission
arising from bremsstrahlung is approximately proportional to
the integral of the square of n_e times the square root of T.
Since the x-ray spectral data by themselves yield T, the
combination of all quantities should give separate estimates of
n_e and the absolute length-scale of the cluster, independent
of any assumed value for the distance.

During the 1970's, the S-Z effect was almost certainly detected in at least a few clusters; however, to make it a practical tool, the accuracy and spatial resolution of the radio measurements must be increased by at least a factor of 10. At present, atmospheric variations and ground pickup appear to set the limits to the attainable accuracy. Sensitivity could be increased somewhat through use of low-noise, broad-bandwidth receivers on large ground-based millimeter-wave dishes. However, the ultimate limits will be achieved only with a space antenna--hence the earlier suggestion for a 30-m space radio antenna to measure these small-scale inhomogeneities in the microwave background. X-ray data having high spatial resolution and moderate spectral resolution are also essential here.

If the x-ray emission is spherically symmetric, comparison of the absolute length scale with the angular size of the emission yields the actual distance to the cluster. This is why the method has been mentioned as one fundamentally new approach to the measurement of the Hubble constant, $\underline{H}_0$, and perhaps ultimately even $\underline{q}_0$ (see Section III). We are currently still a long way from having radio or x-ray data accurate enough to carry out such a program, but it is possible that the microwave measurements and x-ray studies, if pursued vigorously, may eventually have a significance that was quite unsuspected only a few years ago.

3. Other Radio Studies of Clusters

Detailed maps of radio sources in clusters of galaxies will continue to provide a useful probe of the cluster IGM. Most of this work can be done with the Very Large Array (VLA). However, for estimates of confinement pressures, total source energies in magnetic fields and particles must be known. For the power-law slopes that are typically observed in extended radio sources, the total energy is determined by the low-frequency cutoff of the radiation. Observations at low frequencies are therefore required to discover where this cutoff occurs. As a first step, the frequency coverage of the VLA could be extended below its present value of 1.4 GHz to as low as 100 MHz. For some sources, this might suffice; eventually, however, a space synthesis radio telescope operating below 100 MHz should be considered. This facility would also be important in studying the diffuse low-frequency radio halos suspected in some clusters, several of which also show a dramatic increase in flux below 50 MHz. (Further applications of a low-frequency synthesis array are mentioned later.)

4. Superclusters of Galaxies

It is important to verify the recent tentative detections of x-ray emission from superclusters since the amount of mass

involved would dominate the gravitational potential of the
supercluster. To map these extremely diffuse sources of low
surface brightness, we will require x-ray detectors having
higher sensitivity plus moderate angular resolution to
eliminate any discrete background components.

5. Ultraviolet and X-Ray Absorption-Line Spectroscopy

Absorption-line spectroscopy is by far the most sensitive way
to detect a diffuse IGM outside of clusters that is not fully
ionized, since resonance lines become optically thick at column
densities of only 10^{13} atoms/cm^2 in the UV and 10^{15-16}
atoms/cm^2 in the x-ray regions. If the IGM is uniform, we
expect an absorption trough for frequencies greater than that
at which the line is observed in emission in the background
source (the Gunn-Peterson test). If the IGM is clumpy, we
should see absorption lines.

A search for absorption arising from the 304-Å line of
He II in quasars with a red shift of 3 or greater using Space
Telescope (ST) would set stringent limits on the ionization of
the IGM. Metal-enriched regions of the IGM could also be
detected by x-ray observations with high spectral resolution;
the various ionization states of oxygen have a complex of lines
near 0.6 keV that should be observable if the temperature is
3×10^6 K or less. The corresponding complex for iron
appears at 6-7 keV for a temperature of 3×10^7 K or less.

6. Distortions in the Cosmic Microwave Background

As noted above, this test detects absorption below 1 mm by
Compton scattering and/or excess emission due to thermal
bremsstrahlung at wavelengths beyond 20 cm by a hot IGM at high
red shift; the necessary facilities were discussed in Section
III.

7. The Diffuse X-Ray Background

The challenge here is to discover what fraction of the general
background arises from discrete sources, principally quasars.
X-ray observations with high resolution and sensitivity are
crucial. If possible, imaging observations from 10 to 100 keV
might also help because the relative contributions of point
sources and diffuse background could differ at higher energies.

8. The Diffuse Gamma-Ray Background

As in the case of the x-ray background, we need to determine
what fraction of the general background arises in discrete
sources. As mentioned earlier, the spectrum of a truly diffuse
component could tell us whether we live in a baryon-symmetric

Universe with equal amounts of matter and antimatter.
Gamma-ray observations with moderate angular resolution and
sensitivity at 0.5 MeV and below are required.

9. The Far-Ultraviolet Background

Red-shifted Lyman-alpha and He II 304-Å lines are expected to
dominate the far UV emission from the IGM, provided the medium
spent much time in the temperature range 10^4-10^6 K. High
spatial resolution might be needed to eliminate discrete
sources such as stars, galaxies, and quasars; moderate spectral
resolution is needed to recognize individual sources. Because
of the confusion problem, however, a definite detection
strategy has yet to be developed.

VII. QUASARS AND RELATED ACTIVE GALAXIES

A. Review and Summary

We can include only the high points of discoveries made over
the past 10 years in this rapidly advancing field.

Most importantly, the long-standing controversy over the
nature of the red shifts came close to resolution. A number of
low-red-shift quasars were found to reside in groups of galaxies
having essentially the same red shift. Furthermore, some
BL Lac objects, thought to be closely related to quasars, were
also found to be embedded in galaxies, again with the same red
shift. Both of these discoveries strongly support the hypothesis that the red shifts of at least some quasars are cosmological. It seems likely that quasars as a class really are very
distant and hence have extraordinarily high luminosities.

In several radio galaxies, very-long-baseline interferometry
(VLBI) observations showed that small-scale radio structure on
the size of parsecs is aligned with the large-scale radio lobes
having sizes of up to several megaparsecs. This alignment
strongly suggests that the central energy source in the nucleus
has an orientation that remains constant over long periods of
time; quite possibly, the central source is rapidly rotating.
This observation also established a causal connection between
the central energy source and the extended radio lobes, and it
is now thought that energy is supplied to the outer lobes by
relativisitic beams of some sort.

Direct evidence for such beams came from radio observations
of beamlike structures connecting the nucleus with the outer
lobes and also from x-ray observations of a beam in the nearby
radio galaxy Cen A. Observations of small radio "hot spots" in
the lobes of extended radio sources and possible optical emission from these regions are consistent with this idea as well.

Apparent expansion velocities greater than the speed of light were found with VLBI techniques in several quasars and in one active galaxy. The radio structure shows a one-sided jet in each case. This asymmetry provides strong evidence for relativistic motions in these objects.

Recent x-ray observations established that quasars and active galaxies constitute a class of luminous x-ray emitters. This class includes quasars, BL Lac objects, N galaxies, Seyfert type I galaxies, narrow emission-line galaxies, and radio galaxies. Luminosities in the 2-10-keV band range from 10^{41} to 10^{47} ergs/sec.

Direct imaging of the diffuse x-ray background indicated that a large fraction or possibly all of the background at a few keV is composed of discrete sources, mainly quasars.

The quasar 3C 273 was found to have a gamma-ray luminosity comparable with its optical and x-ray luminosities. Gamma-ray emission has also been reported from the Seyfert galaxy NC 4151 and from Centaurus A. Since in all three cases the emission per decade of energy may peak in the gamma-ray range, a new and direct way of studying high-energy processes in at least some active galaxies may now exist. Other quasars and active galaxies were found to emit most of their energy at infrared wavelengths.

Rapid time variability indicating small physical sizes was established for a number of quasars and active galaxies. At optical and infrared wavelengths, variations on a time scale of one day were reported, and polarization in certain BL Lac objects and violently variable quasars was seen to vary on a time scale of hours. At x-ray wavelengths, there have been suggestions of variability on time scales as short as 1000 sec. All of these observations suggest that the central energy source is only 10^{14}-10^{15} cm in size. Since an object's mass is limited by its gravitational (Schwarzschild) radius, the masses must therefore be smaller than 10^9-10^{10} solar masses.

Observations at low radio frequencies (a few hundred megahertz) indicated variations in some sources that are too rapid to be understood in terms of conventional models, but rather seem to require a coherent radiation mechanism or relativistic motions in the line of sight.

The structure of double-lobed radio sources was found to be correlated with their luminosity. For luminous sources, the outer extremities of the radio lobes have the highest surface brightness. For fainter sources, the lobes are brightest near the galaxy and tend to fade out gradually at large radii. Grossly speaking, this correlation might result if the more luminous sources are moving supersonically with respect to the external medium, while the weaker sources are primarily subsonic and are thermally confined.

Among radio sources, the existence of a compact central object associated with a very luminous extended source appears to signify a quasi-stellar object. Roughly 90 percent of the 3CR quasars have nuclear sources, whereas, so far, none of the sources associated with blank fields or distant galaxies has a nuclear component. Since in virtually all cases the nuclear component contributes significantly to the flux density at 178 MHz, there appears to be a close physical relationship between the overall radio and optical emission.

During the last decade it was also discovered that violently variable optical and radio quasars differ from normal quasars in having unusually flat spectra and high fluxes at millimeter wavelengths. The shapes of the spectra in the infrared region and the observed constancy of the polarization vector at all observed wavelengths shortward of a few centimeters both indicate that the regions producing the millimeter-wave and optical radiation are identical or are at least closely related.

Studies of both optical- and radio-selected quasars showed that quasars were much more numerous and possibly more luminous in the past. However, a decline in detected quasars sets in beyond a red shift of 3, and the most distant quasar known has a red shift of 3.5. Limits placed on distant quasars by the x-ray background suggest that this apparent decline in numbers is real and that the increase in quasar numbers does not continue beyond z = 3.

Recent studies have tended to confirm the idea that the mysterious quasar absorption lines have multiple origins. Some evidently arise in the quasar, some in a surrounding galaxy, some in intervening galaxies, and some perhaps in intergalactic clouds. These lines therefore provide a unique probe of gas densities and abundances under a wide variety of conditions at very large red shifts.

Much evidence from the past decade strengthened the earlier notion that there exists a wide and interrelated continuum of galaxies of which quasars constitute only the most extreme examples. Other members of this family include BL Lac objects, Seyfert galaxies of types I and II, N galaxies, and broad- and narrow-lined radio galaxies. However, there are still few firm conclusions as to how the members of this family differ from each other structurally.

On the theoretical side, the most dramatic success was the conclusion that beams of particles or low-frequency radiation emanating from galactic nuclei are possible and can energize extended radio sources. Direct evidence for such beams was obtained in several cases, as noted above.

Among the many possible models for quasars and active galactic nuclei, accretion of material onto a central black hole of mass between 10^6 and 10^{10} solar masses now appears to be favored. Regardless of whether this specific model eventually proves to be correct, however, a still more important conclusion has emerged. Despite the enormous energies involved in some of the outbursts observed in quasars and active galaxies, it has been possible to devise dynamically self-consistent models of these outbursts. There is thus at present no strong theoretical reason to doubt the cosmological nature of the observed red shifts or to believe that "new physics" is required to understand these objects.

B. Prospects for the 1980's

1. The Nature of the Central Energy Source

There appear to be two basic kinds of measurements that can be made over the next decade that have direct bearing on the nature of the central source.

a. The Complete Spectral-Energy Distribution and Its Behavior with Time

The continuous spectra of quasars and active galaxies hold the key to determining the energy-emission mechanisms in these objects. In the radio region it seems likely that most of the emission from strong sources is synchrotron emission, but it remains unclear what the particle distributions and geometry are and whether relativistic bulk velocities are important. It is also not known why most quasars are radio-quiet. In the optical and infrared regions, synchrotron radiation is most often assumed because of the high polarization in some sources, but inverse-Compton scattering or emission by warm dust might prove more important in some objects.

A prime objective will be to observe quasar flares over a broad spectral range (radio to gamma-ray), to map the flares with VLBI, and to follow the development of each flare in time. With very high-resolution VLBI observations yielding a spatial positions of 0.1 pc or better, it should be possible to measure the exact locations of recurring flares. The black-hole model would of course predict a precisely constant location, whereas a dispersed model based on recurrent supernovae, for example, might produce flares in slightly varying locations. Such high-resolution observations would also be crucial in determining the nature of the superluminal expansions associated with some outbursts.

Judging from the preliminary x-ray data now becoming available, x-ray variability appears to be particularly useful in estimating sizes of emitting regions. The small sizes inferred from rapid x-ray variability suggest that, in x rays, we are seeing much more deeply into the source than is the case at radio wavelengths. Simultaneous observations at radio, optical, and x-ray wavelengths will pose a strong test of models. For example, flares that show energy-dependent rise times may indicate an inverse-Compton mechanism, whereas substantial polarization at all wavelengths (including x rays) would favor synchrotron emission. High-energy fluxes above 10 keV are needed to establish overall spectral characteristics and also to set total luminosities. High-sensitivity x-ray spectroscopy may also be able to detect absorption or emission features, which can provide information about gas in or around the x-ray emitting region, as well as measure distances from x-ray red shifts directly.

The observational facilities necessary to carry out this program encompass virtually the entire range of wavelengths used by astronomers. At the lowest frequencies, radio and millimeter-wave astronomy mainly need continued support of existing instruments and programs. Arecibo, the VLA, and the proposed millimeter-wave telescope should all be adequate instruments. Other smaller telescopes can contribute to variability studies on brighter objects. Between 1 mm and 10 μm, space facilities are vital to bridge the apparent spectral discontinuity seen in most sources between the IR and millimeter wavelengths. In the optical wavelengths, short-time-scale variations in polarization are currently flux-limited, and larger collecting areas for ground-based optical telescopes are essential. In the UV, ST appears to be adequate for the present. Of special concern, however, is the hiatus in x-ray observations that began when the Einstein x-ray observatory ceased operations. Future space facilities for x rays should include a telescope with high sensitivity in order to detect rapid variability, polarimeters to test for synchrotron emission, and high- and moderate-resolution spectrometers to study emission and absorption lines. Facilities sensitive to energies about 10 keV (high-energy x-ray and gamma-ray) are important to complete the spectral coverage, especially in view of the large gamma-ray flux from 3C 273.

Most importantly, we urge the planning of coordinated programs to ensure thorough spectral and temporal coverage of at least a few carefully chosen objects. A relatively small number of such observations is worth much more than hundreds of observations of many different objects made at random wavelengths and times.

b. High-Resolution Imaging

Direct observation of the geometry of the central energy source requires high spatial resolution. This need is best filled at present by VLBI, which can actually resolve structure in many sources. The angular resolution possible with VLBI at wavelengths from 1 to 50 cm is currently set by the Earth's diameter. As applied to existing sources, this limit on angular resolution works out to a surface brightness temperature of 10^{12} K. Unfortunately 10^{12} K coincides with the limit that inverse-Compton scattering should impose on the synchrotron process. Brightness temperatures in excess of 10^{12} K would indicate either stimulated emission, a coherent radiation mechanism, or relativistic mass motions in the line of sight. The brightness temperature is thus a powerful clue to source structure. One of the most fundamental pieces of evidence that VLBI can provide can therefore not be obtained from the Earth's surface and requires an antenna in space.

Recent VLBI maps suggest an exciting new development: the inner radio nuclei appear to show jets similar in structure to the large jets mentioned above. In some cases these inner jets appear bent. We do not understand how such jets could be form- ed or how they could remain stable long enough to carry energy from scales as small as parsecs out to distances of hundreds of kiloparsecs. The bending could be due to hydromagnetic insta- bilities or perhaps to a precession of the region producing the jet. VLBI observations clearly appear to contain important information on both the structure of the energy source and the mechanism of energy transport to the outer structure.

Although there seem to be no true quasars close to our Galaxy, there are many active galactic nuclei relatively near- by. Whereas the typical resolving power of VLBI on a quasar might be a parsec, for active nuclei like Cen A or M87 it is as small as a few hundredths of a parsec. The nucleus of even our own Galaxy may be active at a low level, and millimeter-wave VLBI observations in this case have the potential to resolve structure down to 1 AU. Continued VLBI studies of these near- by active nuclei should provide models for the inner nuclear structure in quasars as well.

Present facilities for VLBI observations need improvement. A practical problem has been the incomplete coverage in the Fourier-transform plane that is possible with existing VLBI telescopes on Earth; this situation could be improved by in- creasing the number of antennas. Another approach is to orbit an antenna in space and use the rapidly changing baselines with Earth-based telescopes to improve the coverage. A space an- tenna in a highly eccentric orbit could also push beyond the 10^{12} K brightness temperature limit, as noted above. Which- ever approach is taken, VLBI in the 1980's seems ripe for a major advances.

With its much shorter working wavelength, optical interfer- ometry offers a potential gain of 10^4 over radio techniques for receivers operating on baselines of the same size. Ground- based optical telescopes do not usually exploit this potential resolving power because of blurring by atmospheric seeing. Although techniques have been devised to overcome this degra- dation (e.g., speckle interferometry), they can be applied at present only to objects brighter than 15th magnitude, even with

large telescopes. We are not certain at this time whether
ground-based or space-based approaches to optical interferometry will prove to be more fruitful; however, it is clear that
much of the information potentially available in the optical
and infrared spectral regions is currently not being recover-
ed. The 1980's are an opportune time to begin serious design
studies for optical interferometers, followed by prototype
construction. As an ultimate goal, we should advance our
knowledge of interferometric techniques to the point where
construction of an instrument capable of attaining a resolution
of 10^{-4} arcsec down to $\underline{M}_V = 19\text{-}20$ can be contemplated in
the 1990's.

2. Luminosity Functions, Number Densities, Evolution with Red Shifts; the Nature of the Diffuse X-Ray Background

Indirect evidence about the nature of quasars and active
galaxies can be obtained from their luminosity functions at
various wavelengths and from evolution in these functions with
red shift. This problem is being extensively studied at radio
and optical wavelengths. X-ray observations provide a new and
powerful method for locating quasars and active galaxies,
virtually all of which appear to be strong x-ray emitters.
Furthermore, the confusion due to background sources is a much
less serious problem, since the areal density of x-ray sources
is many orders of magnitude smaller than the density of optical
sources. X-ray observations should be capable of detecting
quasars at red shifts of 4-10, if such objects exist. Deeper
x-ray surveys will also help to establish what fraction of the
x-ray background arises in discrete sources, as noted previously.

Ground-based optical surveys for quasars have recently
revealed hints of clumping. At present, this is our only
evidence for density inhomogeneities at early epochs in the
Universe. Further surveys of this kind carried out from space
are strongly encouraged because freedom from atmospheric seeing
will ensure uniform magnitude limits in all fields. Lyman-
alpha, the primary signature for quasars, will also be
detectable at all wavelengths, eliminating the worry that the
detection frequency versus red shift is subtly modulated by
atmospheric transmission or sky brightness. Despite the rather
small field of view of ST, initial space surveys should be made
with this instrument since the anticipated population of
objects is large. To study clumping on larger angular scales,
however, a space survey instrument with a wider field of view
may eventually be necessary.

3. The Physics of the Energy Ejection Process; the Nature of
Extended Radio Sources

The complex nature of radio jets and lobes is only poorly
understood at this time. Jets appear to offer evidence of
energy transport from the nucleus to the outer structure. A
complete description of jets would include the relativistic-
particle density, the magnetic-field strength and configuration, the thermal-particle density, and the macroscopic
velocity field.

Polarization and intensity maps made with the VLA from 1.4
to 22 GHz will supply most of these quantities, provided
certain important model parameters are known. The low-energy
particle cutoff must be measured, since this usually fixes the
total energy in radiating particles. This lower cutoff has
generally been assumed to be 10 MHz, simply because that is the
cutoff frequency of the Earth's ionosphere. For this reason,
as noted earlier, an important area of radio astronomy during
the 1980's should be the improvement in mapping capabilities
below 1 GHz. The range of the VLA should first be extended to
lower frequencies (100 MHz), but ultimately a facility above
the ionosphere should be considered.

The spectral cutoff at high energies is also important, as
it fixes the decay time of the radiating electrons and hence
the lifetime of the source. Measurement of this cutoff in a
given source will often require millimeter-wave observations.
During the next decade, the two 25-m millimeter-wave telescopes
proposed by the United States and West Germany should make
progress here.

Both optical and x-ray studies of the emission from jets
and lobes are of growing interest. Optical emission from one
hot spot in a radio lobe has been detected but suggests a surprisingly cool gas at about 10^4 K. At least four optical
jets are now known, in addition to the x-ray jet in Cen A.
Radiation at such high frequencies implies very short lifetimes
for the high-energy electrons, which must somehow either stream
directly along the field lines or be accelerated in situ.

X-ray observations of extended radio lobes should allow a
separate determination of the density of high-energy electrons,
with Compton-scatter from the microwave background. When
combined with maps of the radio intensity, polarization, and
spectral index, this information should permit one to deduce
the particle density and the magnetic-field strength and
configuration in the radio source.

These observations require the same instrumentation needed
for other studies. ST should revolutionize the optical imaging
of jets and hot spots in lobes. X-ray observations with high
spatial resolution and high sensitivity are needed for radio
lobes. Imaging at energies above 10 keV, if possible, would
help to separate thermal emission from the nonthermal Compton

process. It is also essential to understand the theoretical problems of the propagation of relativistic particles through a magnetized plasma.

4. The Relation between Active Nuclei and the Galaxies and Groups of Galaxies in Which They Are Embedded

Because it is expected to produce sharp images of galaxies associated with relatively nearby quasars, ST should be our single most important instrument for studying the relationship between galaxies and quasars. The first priority here is simply to test the basic hypothesis that quasars are superactive galactic nuclei. Assuming that answer to be affirmative, we will then need to study how the presence of the active nucleus or quasar depends on the morphological type, mass, and environment of the ambient galaxy. Several puzzling facts are already known about such correlations, but at the moment they seem to point to contradictory conclusions. At low red shifts, for example, quasars are found only in sparse clusters, which are typically populated by spiral galaxies; this fact suggests that quasars may be more closely related to Seyfert nuclei, which are also found only in spirals. On the other hand, when quasars are radio sources they show the classic double-lobed structure typical of giant radio galaxies, which are invariably elliptical galaxies. How does all of this evidence fit together? This brings us directly to our next point.

5. The Relationship between Active Nuclei and Apparently Normal Nuclei

The main question here is what factors induce activity in a nucleus. Are they to be found primarily within the nucleus itself, or is activity stimulated by the right combination of environmental influences? The relationship between the active nucleus and its ambient galaxy discussed above is only one side of this coin. The other side will come from studying the properties of nearby, ostensibly normal nuclei. Could it be that dormant quasars are lurking within many local galaxies, currently inactive because their fuel supply has been exhausted after billions of years?

An important instrument here is again ST, which will be able to place much better limits on the mass and density of material in the very centers of local galactic nuclei. The direct-imaging capabilities of ST will be adequate as currently planned. However, the telescope urgently needs a two-dimensional detector for spectroscopy that is capable of exploiting the high angular resolution of ST along the slit. This is needed to measure rotation curves and velocity dispersions as close as possible to the cores of nearby galactic nuclei. This instrument could perhaps be combined

with the high-resolution ultraviolet echelle spectrograph
mentioned in the previous section.

Sensitive radio and x-ray observations of large numbers of
normal and active galactic nuclei will reveal various levels of
nuclear activity. Luminosities coupled with time scales for
variability can be used to constrain sizes of emitting regions
and estimate masses of possible collapsed central objects.

There are also many theoretical questions to be investigated concerning the creation and refueling of active nuclei.
Strong gravitational torques, as in bars, or gas accretion
during mergers with other galaxies are just two of several
possible mechanisms that might induce a rapid infall of gas
into the nuclear regions, perhaps thereby triggering the
fuel-starved nuclear source.

6. Surrounding, Ejected, and Intervening Material

The gas producing the prominent optical emission lines seems in
many cases to be only about a light-year from the central
energy source. An understanding of its dynamics and other
physical conditions could tell us much about the central source
itself. ST will provide broadened spectral coverage at short
wavelengths. Infrared spectroscopy is needed for observations
of forbidden lines in high-red-shift quasars, which will yield
valuable information on the density, temperature, and dust
content of the material surrounding the quasar. For complete
spectral coverage, these measurements should best be made from
space at wavelengths of 2.5 µm or greater.

A rich literature has sprung up on the absorption-line
components in quasar spectra. Recent statistical analyses of
various types of lines suggest that some of these absorption
lines originate in the quasar itself, some in the surrounding
galaxy, and others in intervening galaxies or extragalactic
clouds. The evolution in the space density of these intervening clouds (as ascertained from their frequency distribution
with red shift) can tell us much about the evolution of
clustering in the Universe and how and when gas accreted into
galaxies.

Further study of these clouds is today hampered by inadequate spectral resolution. The clouds are so numerous that the
individual components are unresolved in most quasars. Higher
resolution requires larger ground-based telescopes for more
light.

Another problem is that most of the absorption lines are
Lyman-alpha lines and are therefore visible from the ground
only in highly red-shifted quasars. ST could provide needed
wavelength coverage in the UV region, but a high-resolution UV
spectrograph with ample spectral range is required. The
two-dimensional UV echelle spectrograph mentioned above could
meet this need. A resolution on the order of 10,000 would be

appropriate, with continuous spectral coverage from 1200 to
3200 Å in a single exposure.

VIII. GENERAL REMARKS

We conclude this chapter with a few additional matters that
could not be covered adequately in the preceding sections. The
first point concerns the desirability of a large computer designed especially for N-body computations. Such a computer
could be extraordinarily useful in several important theoretical studies, including those of gravitational clustering,
galaxy collisions and mergers, galaxy collapse, and the evolution of star clusters. We support the idea of such a machine,
provided it would furnish a significant increase in capability
over computers now available--that is, if it could accommodate programs involving a minimum of 100,000 to 1,000,000
particles. However, we also believe that progress on most of
these problems can still be made with facilities available at
typical university centers and with the increasingly powerful
minicomputers; in view of the convenience and flexibility of
such computers, continued support for these facilities must
have higher priority. Furthermore, a huge computer cannot
substitute for the imagination, insight, and expertise of
well-trained theoreticians. In short, a large N-body computer
could be highly beneficial, but not if it entailed a significant decrease in existing levels of support for theoretical
astrophysics.

Our review of extragalactic astronomy during the past 10
years demonstrates that rapid advances are occurring on all
wavelength fronts simultaneously and that the development of
wavelength bands outside the traditional optical window has
been essential to the present vigor of the field. At the same
time, it is clear that virtually every new discovery in other
wavelength ranges has placed new demands on existing optical
facilities because some sort of optical measurement is usually
required to follow up the original observation. As a typical
example, suppose a new, variable x-ray source is found. An
excellent photograph of the region is required to identify
possible optical counterparts. Potential candidates are
located; they are usually quite faint. Time-consuming optical
spectroscopy is then necessary to ascertain the nature of the
object--whether QSO, distant Seyfert galaxy, or star--and to
determine its red shift if necessary. Thus the original x-ray
discovery, which might have taken only an hour or so for each
typical x-ray source, eventually requires many hours of
optical-telescope time to be fully understood. The tragedy is,
of course, that if these optical observations cannot be made
soon enough to catch the object in its misbehavior, or soon
enough to suggest follow-up studies to the x-ray observers, the

full benefit of the highly expensive x-ray (or radio or UV) observations will be lost.

This problem is exacerbated by the fact that there are essentially only two large telescopes--the 4-m instruments at Kitt Peak National Observatory (KPNO) and at the Cerro Tololo Inter-American Observatory (CTIO)--that are openly available to American x-ray and radio astronomers for optical backup measurements, and these instruments are both heavily oversubscribed. As a quantitative example, consider a recent collection of dark-time proposals for a 6-month period on the KPNO 4-m telescope. There were 71 requests, of which the great majority were of high quality and clearly deserved to be granted. (Sixty-six of the proposals dealt with extragalactic astronomy, an overwhelming testimony to the present vitality of our field.) Over half of all proposals expressed a need to follow up on discoveries made with x-ray, radio, and ultraviolet satellite telescopes. Unfortunately, since a total of only 20 dark-time requests could be accommodated on the 4-m telescope during this 6-month period, most of these follow-up projects could not be carried out.

Evidently the opening up of extragalactic astronomy to other wavelength bands, far from decreasing the interest in optical observations, has actually increased it. ST may be expected to increase the pressure still more, for its relatively small aperture cannot provide the large light-gathering capacity necessary for optical spectroscopy, which is still our primary tool for determining the nature of astronomical sources and for measuring their red shifts.

To resolve the optical logjam, we may follow two avenues: increase the number of optical telescopes, or increase the efficiency of existing instruments. With regard to the former, the Working Group did not attempt to decide the number of telescopes of a particular size that would be optimal. However, we note that the demand for the smaller telescopes at KPNO and CTIO (2.1 m and smaller) is not so severe as that for each 4-m. Thus, to be truly helpful, additional optical telescopes should be of the 3-m class or larger.

The other path open to us--increasing the efficiency of existing telescopes--is extraordinarily cost-effective. Detector efficiencies are now nearing their theoretical maximum values as a result of the intensive development over the last decade of charge-coupled devices and other high-efficiency detectors. We must now take vigorous steps to disseminate these detectors as widely as possible throughout the astronomical community.

It would also be highly desirable to supply funds to ensure that the development of new detector chips is properly matched to optical telescopes. The main problem here is field size: present, high-efficiency detectors are not large enough to cover the image plane of a typical telescope or spectrograph.

Since, for many extragalactic problems, the telescope time required is inversely proportional to field covered, the development of larger detectors is a straightforward way to increase telescope efficiency. Next-generation telescopes will have even larger field sizes, making the problem still more critical in the future.

In summary, the goal of securing adequate access to optical observations for astronomers active in all wavelength bands should have high overall priority for the coming decade.

5

Related Areas of Science

I. INTRODUCTION

The Working Group on Related Areas of Science was charged with
examining the areas of science at the boundaries of astrophysics
and their likely interactions with astrophysics during the
coming decade, as radically new directions are as likely to
come from without as from within. We have not ventured into
technology--detectors and the like fell within the purview of
the Panels--but rather have considered only scientific interactions with other fields.

We have almost certainly been too strongly guided by
hindsight. Only a few years ago, the thought that solid-state
physics or even elementary-particle physics would have much to
do with the mainstream of astrophysics would have seemed
unlikely at best. It is worth remembering too, that the ways
we came to be interested in those fields today are quite
diverse: solid state and condensed matter, through the astronomical discovery of pulsars; and elementary-particle physics,
on the other hand, through the development in that field of
powerful gauge theories that are clarifying old questions
(e.g., neutrino-powered supernovae) and are beginning to allow
us to ask new ones that heretofore seemed unanswerable. But
hindsight is clear; the reader will therefore see represented
here fields already influential in astrophysical thought and
may be alarmed by the absence of topics that simply did not
occur to the Working Group--or, more precisely, to its Chairman.

It must therefore be borne in mind that the report of this
Working Group is not complete, and cannot be. Most of the excitement in astronomy has come from discoveries that could not

have been predicted; there is no reason to expect that things will be different in the decade to come. One of the great dangers of a decade review is that it can too strongly prescribe the development of the science; the field (and this, of course, includes the funding agencies) must remain flexible enough to deal with the unexpected. The very richness of the present interactions with other disciplines and the vast technological innovations occurring now in astrophysics make it even more certain that the unexpected will occur. We should be ready.

This chapter is divided into sections more or less in accord with the contributions from the individual Working Group members, but the Chairman has taken the liberty to rearrange and change emphasis extensively where appropriate.

II. FLUID MECHANICS

Contemporary problems in astrophysics are coming more and more to demand the knowledge and skills of fluid mechanics for their understanding and solution. Stellar pulsation, cloud collapse, a plethora of infall phenomena, accretion disks, binary mass transfer--the list of problems that have called on its techniques during the past decade is long indeed. Recent developments in fluid dynamics promise to be of great utility in the solution of related astrophysical problems, and foreseeable directions in astrophysics will continue to demand developments in fluid dynamics.

A research tool that has seen vastly increased use in fluid problems in the recent past is numerical nonlinear modeling. Some use has been made of it in astrophysics, but the increased availability of large, fast computers and the desire for more reliable and accurate models for problems than were possible in the past will surely bring the technique into increasing use for astrophysical problems in the coming decade. The complexity of astrophysical situations will often preclude a completely realistic model, but much deeper insight will often be gained from detailed modeling of simpler problems than can be provided by qualitative analytic arguments.

Particularly amenable to such techniques are problems related to turbulence, for which classical mixing-length arguments are more and more recognized as being inadequate for understanding phenomena in the detail often demanded by otherwise sophisticated models. The energy flux carried by convection, convective overshoot, hydromagnetic dynamos, and turbulence in radiation-pressure dominated fluids are all areas in which considerable progress may be expected. It is also clearly of importance for astrophysicists to persuade fluid dynamicists to incorporate more physics into their theories, if only to broaden their applicability to the usually extreme conditions relevant to astrophysical problems.

Related areas in which considerable emphasis is being placed in fluid dynamics is the study of intermittency and of nonlinear waves, both of which have immediate applicability to astrophysics through the physics of accretion disks, chromospheres and coronas, and perhaps even the origin of spiral structure. Other areas in which progress may be expected are the following:

* Two-phase flows, such as stars with gas in galaxies, dust with gas in protostars, and radiation with gas in hot stars and the young Universe;
* Dynamics of chemical waves, such as oscillating nuclear reactions in stellar cores and coupling to stellar pulsations;
* Coupling of convection and pulsation for application to variable stars;
* Interaction of fluid dynamics with spectral radiation for application to the theory of spectral lines and their changes, and to the propagation of sound waves in an optically thin medium;
* Dynamics of laser fusion, for understanding dynamics with strong radiative forces;
* Theory of drops and bubbles, for application to formation of dust through the liquid-drop phase; and
* Biofluid dynamics--fluids with very high molecular weight may permit simulations of small-scale heights on Earth and would allow the study of the analog-to-compressible convection in the laboratory.

III. PLASMA PHYSICS

In many respects the relationship of plasma physics to astrophysics is similar to that of fluid dynamics, particularly insofar as the prospects for vastly increased use of nonlinear modeling are concerned. The situation is somewhat different, however, in that the outstanding astrophysical problems are in areas in which there is little understanding among practicing plasma physicists, and these will certainly not yield to extensions of laboratory or terrestrially motivated work already in progress.

Perhaps the outstanding example is the problem of magnetic-field reconnection, which has been and remains the elusive key to many (perhaps almost all) high-energy phenomena in astrophysics. Unless it proceeds at the Alfvén rate, as is usually assumed on the basis of little evidence, the magnetic field in differentially rotating astrophysical fluids could be quite different from that usually assumed. Such important but highly uncertain processes as magnetoturbulent viscosity depend crucially on reconnection as well as understanding of compressible magnetohydrodynamic (MHD) turbulence, which will come--if at

all--through numerical modeling. The emphasis so far in MHD
theory has been on laboratory (particularly fusion) plasmas,
and much more research needs to be focused on astrophysical
problems.

Some recent results in plasma physics have applicability to
astrophysics directly, and applications of them will doubtless
be forthcoming in the near future. Two examples are the notion
of stochasticity--the transport of particles in phase or configuration space in violation of the adiabatic invariants by the
action of resonant effects--and the nonlinear limiting of
instabilities, which affect the propagation of energetic beams
such as cosmic rays.

There is perhaps no "related" science of such immediate
applicability to astrophysics that is also marked by such difficulty of communication as plasma physics. It is doubtless
the complexity of the phenomena treated that is primarily
responsible, but it seems clear that both fields would benefit
from closer ties. Possible ways to facilitate this might include the availability of astrophysically oriented texts and a
series of joint conference-workshops. It has also been suggested that an encyclopedia of astrophysical plasma physics,
such as is being prepared for laboratory plasma physics, would
be helpful. In any case, it seems highly desirable to
establish closer ties.

IV. ATOMIC AND MOLECULAR PHYSICS AND CHEMISTRY

The considerations that led to the development of astrophysics
at the turn of this century sprang from atomic spectroscopy.
It is therefore not surprising that atomic physics--and later,
with the consideration of lower temperature processes,
molecular physics--should have remained closely connected with
astrophysics and its problems. Their impact is felt primarily
in problems concerned with relatively low-density regimes, from
the extreme dilution of the intergalactic gas to the densities
found in stellar atmospheres, although in the study of
opacities, conditions of very high density are also encountered.

The classical problem with the application of laboratory
atomic and molecular research to astrophysical problems--that
of duplicating the conditions and species of interest--has if
anything become more acute since astrophysical interest has
broadened to include conditions of extreme dilution, very high
temperatures, and even more transient species than before.
Technological progress in laboratory research has been exceedingly rapid, but the experiments are still difficult and are
becoming increasingly expensive. A recent study for the Astronomical Sciences Division of the National Science Foundation
pointed up the funding crisis that astrophysical applications
of atomic and molecular physics suffer: there are people,

laboratories, and probably sufficient interest in problems crucial to astrophysics, but there is little money to pursue them. Another problem is one of the same ones suffered by plasma physics--namely, that astrophysically related research is of greater interest to astrophysicists than to the atomic/ molecular researchers' peers. The lines of communication in the past have been good; they are perhaps now not so good and in any case are in need of improvement.

The general areas in which research in atomic/molecular physics and chemistry can be expected to affect the development of astrophysics in the coming decade include the following:

* Electron-atom collisions, including excitation and dielectronic recombination;
* Electron-molecule collisions, primarily dissociative recombination;
* Atom-atom collisions, primarily charge transfer;
* Chemical reactions, including surface reactions, radiative association, and molecular exchange reactions;
* Inelastic atom-molecule and molecule-molecule collisions; and
* Photo processes, including photodissociation, photo-ionization, photoexcitation, and photoprocesses in strong magnetic fields.

As mentioned above, laboratory technology in the last decade has advanced enormously and is still doing so, primarily but not exclusively through the development of tunable lasers, which can soon be expected to cover the range from the near infrared through the energies relevant to the ground-state electronic excitations of most of the abundant species in astronomy. Other important developments have been in the pro-duction of tightly controlled beams--from very cold, slow ones to high-current, highly ionized ones--and in the construction and use of high-intensity synchrotron continuum sources for the UV and x-ray regions. A region of great interest to astro-physics that has been neglected in the laboratory is the region from 50 to 1000 μm, which contains important vibrational and rotational transitions pertinent to the interstellar medium. Development of detectors for astronomy will doubtless aid this effort, and exciting results in the coming decade may be hoped for.

The capability for laser techniques to prepare and monitor very short-lived species, which are potentially of revolution-ary interest for laboratory astrophysics, has just begun to be exploited. For example, the "resonance ionization spectros-copy" technique makes it possible to detect even single atoms of a species, which for the first time raises the possibility of studying even extremely reactive and short-lived species and of studying directly extremely slow processes that are fre-quently of interest.

The techniques for producing very slow and/or cold beams, in conjunction or not with laser techniques, are making it possible to study collisional processes at the very low temperatures found in molecular clouds, including vitally important ion-molecule processes.

The interests of high-energy astrophysics can be (and to some extent are being) served by the techniques of plasma fusion research utilizing very high-temperature, highly stripped plasmas. There is also renewed interest in the properties of matter in ultrahigh magnetic fields, from 10^6 to 10^{13} gauss, which is motivated by pulsars and magnetic white dwarfs. Fields at the low end of this range are accessible in the laboratory, and work in this area may be expected to advance in the coming decade.

Lastly, laboratory studies in the surface chemistry relevant to grain-molecule reactions are crucially needed. The achievement of astrophysical conditions in the laboratory is difficult, but this work must be supported.

On the theoretical side, the areas in which recent progress has been made and which will continue to have an impact on astrophysics include several techniques for the more accurate calculation of the physics of many-electron systems and for dealing with collision problems in which many states (usually of angular momentum) are present and relevant. Work in these areas is expected to be facilitated by the advent of the recently established National Resource for Computation in Chemistry, both through the dissemination of the (usually very complex) codes required and through the provision of computer time.

Theory is currently having an impact on an area that is of direct relevance to astronomy but in which there is really no relevant laboratory work, namely, in the calculation of low-energy charge-transfer reaction rates. One can expect that, as computational techniques and facilities develop, accurate calculations for many processes of great interest to astronomy (but difficult or impossible to study in the laboratory) will become possible. The interplay between theory and experiment, crucial in any case to ensure the validity of theoretical results, is especially important in this work because of the vast gulf in physical conditions that must be bridged.

V. NUCLEAR PHYSICS

Traditionally, nuclear physics has been the handmaiden of astrophysics through the study and understanding of nuclear processes in stars and in the low-energy component of cosmic rays; this role can be expected to continue. The "winding down" of low-energy nuclear research in the United States can be viewed only with alarm by the astrophysical community, which is and will continue to be dependent on low-energy laboratory work of ever better quality as better and more accurate models of stars in all stages of evolution are constructed.

There are five areas that will probably greatly affect astrophysics in the coming decade to which we wish to draw special attention: the solar-neutrino question, stellar nucleosynthesis theory, cosmic-ray isotopic composition, nuclear chronology, and supernova modeling.

The solar-neutrino problem remains one of the most perplexing in astrophysics, striking as it does at the very roots of our understanding of stars and stellar evolution. One important result of the last decade is, of course, the probable detection of solar neutrinos by the ^{37}Cl experiment, but still at a level (2.2 $\pm$ 0.4 SNU) much smaller than that predicted by theory. Recent laboratory results on an important cross section [for ^{3}He(alpha,gamma)^{7}Be] and new calculations for the solar opacities have produced large differences (in opposite directions) for the theoretical flux, and it still remains about 2-1/2 times as large as the measured one. It would seem of very high importance to re-examine carefully all the relevant cross sections and atomic-physics calculations that go into the nuclear reactions and stellar models. With the new high-current accelerators, sensitive detectors, improved atomic models, and faster and bigger computers, this is all possible and should have the highest priority.

Stellar-nucleosynthesis research is expected to take a new and qualitatively different turn in the coming decade, thanks to the large increases in computer capability that we have seen and, it is to be hoped, will continue to see. Instead of the idealized, parameterized models of the past, calculations will be carried out in specific astrophysical contexts for specific stars of specific composition. The sensible application of these techniques requires not only large computers but also accurate reaction rates and realistic evolutionary models for stars (all of which again points up the importance of the solar-neutrino problem). Such a program should produce testable predictions for differences in nucleosynthetic products from stars of different metallicity, a reliable framework for interpretation of astronomical abundance data, and information about the production of short-lived nuclides that may carry important evidence on the formation of the solar system and may produce decays that are observable from satellites such as the Gamma Ray Observatory. Most importantly, these calculations may provide a framework for constructing accurate chemical evolutionary models for the Galaxy, including constraints on the behavior of past stellar populations. Also emerging from these studies should be an understanding of the puzzling isotopic anomalies now being found in primitive solar-system material--if only by forcing the admission that postformation processes might be responsible for them.

Studies of cosmic-ray isotopic composition are expected to advance significantly in the next decade with the flight of planned satelliteborne experiments. Isotopic ratios in principle carry a wealth of information about the abundances and nucleosynthetic history of the material at the source--much less ambiguously than do elemental abundances. Recently, numbers of experiments have revealed isotopic ratios that differ significantly from the solar-terrestrial ones, and future work promises to be very exciting. Questions that might be answered include the old one of whether the cosmic rays are accelerated from the material of the interstellar medium or from fresh supernova ejecta; if the latter, what sort of supernova; if the former, what is the metallicity and overall abundance pattern of the region from which the cosmic rays come? For the light elements, of course, one gets information on spallation, which with laboratory data permit correct interpretation of the results for the heavier species.

The great hope for nuclear chronology in the coming decade is better understanding of the potentially extremely powerful Re-Os chronometer, through better understanding of the relevant neutron capture cross sections, thermal corrections to that of ^{187}Os for excitation, a better measurement of the ^{187}Re half-life, and better understanding of the abundances and attendant geochemical effects for rhenium and osmium. Significantly better understanding of the uranium-thorium system will probably also come through the inclusion of more sophisticated nuclear physics and better r-process models.

The important question of how the products of nucleosynthesis are returned to the interstellar medium may be answered some day by the construction of supernova models. The mechanism of the explosion is still unclear, and many promising and exciting effects of the last decade have not significantly improved the situation. The resolution of this problem may well come with accurate two-dimensional models of collapse, in which the effects of rotation, nonlocal convection, and magnetic fields can be included. Improvements in computer technology will facilitate these developments, but much work at the conceptual level remains to be done.

VI. METEORITE STUDIES AND SOLAR-SYSTEM COSMOCHEMISTRY

In the past few years a series of exciting observations has demonstrated that the solar system contains isotopically heterogeneous materials from different stellar sources. Some of these sources appear to have injected freshly synthesized materials formed immediately (about 3×10^6 years) before the solar system formed. This injected material contained short-lived radioactive nuclei and stable nuclei that can be used to establish the state of the interstellar medium from which the solar nebula separated. Current estimates of the hydrogen density in the medium from which the solar system formed based on the ^{26}Al abundance (half-life = 7 x 10^5 years) indicates a value of about 10^3 atoms/cm, suggesting that solar-system material separated from a dense interstellar cloud containing massive, rapidly evolving stars. Demonstration of the existence of isotopic heterogeneities and of extinct short-lived radioactive elements in the solar system has opened up the whole field of solar-system formation. In addition, these observations give insight into the particular nuclear astrophysical processes within supernovae by revealing the isotopic and chemical state of supernova debris and interstellar matter. It now appears that reliable data can be obtained in the laboratory on primitive solar-system materials (meteorites and comets), which furnish views of the early stages of condensation in the solar nebula. These discoveries have given rise to a new generation of experimental and theoretical studies that will expand and define a major field of activity for the next decade, with the participation of diverse groups with interests including nuclear physics, astrophysics, and cosmochemistry. Because it is now possible to analyze directly in the laboratory aggregates from the early solar nebula, interplanetary dust grains, and partially preserved pre-solar-system materials, there will be a major shift in experimental approaches. Many different laboratories will expand their experimental activities to include the study of these exotic materials and will move toward the use of sophisticated equipment and techniques that will allow the study of "grains" made up of 10^{10}-10^{11} atoms. Extended efforts will be undertaken to collect and study the morphology, chemistry, and isotopic composition of interplanetary and dust grains and cometary material. Serious efforts will be initiated to study the mechanisms of grain condensation, grain formation, and plasma effects on Earth laboratories and in space laboratories under conditions simulating the solar nebula and some stellar atmospheres. The distribution of different types of stellar debris in the dust and gas phases of the interstellar medium will come under close scrutiny, and the chemical interaction of these materials during aggregation will be reasonably understood. The laboratory investigations will provide sufficient clues to permit the development of sound theoretical models for the formation of the solar system. In fact, we shall see the growth of a truly multidisciplinary scientific community that is actively working to understand the detailed state of the interstellar medium and the mechanisms of accumulation of matter to form stars.

VII. THE PHYSICS OF HIGHLY CONDENSED MATTER

The behavior of matter under extreme conditions of density, temperature, magnetic, and gravitational fields is one of the frontier problems in theoretical physics. It is not only a challenging problem in itself but also central to our understanding of pulsars, compact x-ray sources (accreting neutron stars, black holes, and white dwarfs), gravitational collapse, and the early Universe. During the past decade, extended temporal and broadband spectroscopic studies carried out by x-ray astronomical satellites have led to the identification of specific compact x-ray sources as accreting neutron stars, black holes, and white dwarfs in close binary systems. Such sources provide a unique opportunity to study matter under extreme conditions not accessible in the terrestrial laboratory. Quantitative theoretical models demonstrate that detailed studies of these sources will lead to a greatly increased understanding of dense and superdense hadron matter, hadron superfluidity, high-temperature plasma in superstrong magnetic fields, and the possible existence of pion condensates, neutron solids, or quark liquids in the cores of neutron stars.

Among the topics related to neutron-star physics of current interest and in which one many expect significant progress in the coming decade are the following:

* The equation of state of cold neutron matter at densities greater than that of nuclear matter, which determines the mass-radius relation, crustal extent, and maximum mass of neutron stars. The problem is difficult both because the interaction between neutrons is not accurately known and because the $\underline{N}$-body problem is an extremely difficult one.

* Neutron and proton superfluidity, which strongly influences the "glitch" behavior in pulsars, and for which those phenomena provide important diagnostics. Of particular interest is the behavior of the inevitable vortex lines in a rotating superfluid star. Several regimes of superfluidity are expected to exist in most neutron stars, including a superconducting proton fluid in the neutron fluid interior.

* Possible new phases of matter at very high densities, including a condensed Bose pion fluid, and/or a phase transition to a quark liquid. Such states might affect both the evolution and dynamic response of massive neutron stars.

* The equation of state of hot (about 10 MeV) high-density matter determines the behavior of collapsing cores, and its understanding is probably crucial for proper understanding of the supernova phenomenon and the production of supernova remnants.

* The behavior of condensed matter in superstrong magnetic fields may well determine the behavior of the pulsar phenomenon through its influence on the conductivity and effective work function of the neutron-star surface.

The growth of high-energy observational astronomy will certainly continue to provide an arena for the interaction of the physics of condensed matter and astrophysics, to the benefit of both.

VIII. RELATIVITY

The interactions between the General Theory of Relativity and astrophysics have always been close, since most of the interesting classical relativistic effects occur on astronomical scales.

The General Theory of Relativity has reached a mature stage, at least as far as its long-range, nonquantum predictions go. The important phenomena of the theory--such as black holes, gravitational waves, and big-bang cosmology--are thought to be fairly well understood, although some significant theoretical issues remain open. On the other hand, the theory has been tested only in weak-field, solar-system experiments. The agreement with these tests is by now quite striking because of major instrumental innovations in the last two decades (very-long-baseline interferometry and spacecraft tracking), but proof of the existence of black holes and gravitational radiation is still lacking and keenly needed. Black holes may be important in some compact x-ray sources; in one class of models they act as the "central machine" in quasars and active galactic nuclei. Gravitational waves, when detected, will provide a new astronomical window. There are even some distant but admirable hopes that quantum gravity, which does not yet exist as a complete theory, could have some observable consequences.

The primary areas of contact in the next decade will probably be mostly in the areas of the astrophysical studies of black holes and in the continued quest for detection of gravitational radiation. The present situation regarding the existence of black holes is well known: several binary x-ray sources provide plausible but not absolutely convincing candidates, and there are dynamical and photometric data for the nuclei of a few galaxies implying that large central mass concentrations exist--again, plausibly, black holes. The search for further candidates and further elucidation of the nature of currently known ones will come naturally from present studies in observational astrophysics in all wavelength regions. The theoretical support will come not from relativity theory but from fluid dynamics and radiation physics.

The proof of the existence of and/or the detection of gravitational waves is expected to be a major area for study in the next decade. Improvements in resonant-bar detectors, laser interferometers, and precision spacecraft tracking will all be important in different frequency regions; the second will conceivably achieve sensitivities of about 10^{-21} in the

dimensionless strain within the decade, allowing detection of
nonspherical collapse of a star as distant as the Virgo
cluster. It is important that the capability for suitably
precise spacecraft tracking be built into the probe missions;
the incremental cost is small and the scientific return possibly very high.

Observations of the effects of radiation reaction (such as
have recently been found for the binary pulsar) are also
important but do not have the same force as a direct detection
of the waves themselves, which carry attendant information on
transversality and polarization. One can at this time only
speculate on the astrophysical implications of the detection of
gravitational waves, but it seems likely that if the collapse
of stars generates them at all, they will be an important probe
of the collapse phenomenon.

Theoretical work of interest to astrophysics is likely to
concentrate on the short-range quantum regions of General
Relativity, for which no satisfactory theory yet exists. It is
conceivable that the next decade will bring a satisfactory
unified quantum theory of all forces, including gravity, which
would doubtless have great impact on our ideas of the physics
of the early Universe. A possible macroscopic prediction of
such a theory might be the existence of a nonzero cosmological
constant or even (conceivably) a small scalar part of the
gravitational field.

Quantum effects can also appear in the physics of the explosion through the Hawking mechanism of mini black holes; if
these are produced in the early Universe, their existence would
provide important clues to the conditions then, and their
demise now would provide data on elementary-particle spectra
and interactions at energies far beyond the reach of laboratory
techniques. Searches in the gamma-ray and radio bands for
these explosions may well be justified, albeit as a small-
scale, high-risk venture.

Another area in which there will certainly be work with
more prospects for progress is the origin of the fluctuations
from which galaxies and clusters form. The answer to this
question may again come through elementary-particle physics.

IX. ELEMENTARY-PARTICLE PHYSICS

Until recently, the interactions between particle physics and
astrophysics were essentially nonexistent, but developments in
both fields have brought them much closer--so much so now that
perhaps the most exciting problems in all of astrophysics (and
perhaps even in particle physics) lie in the overlap between
the two disciplines.

The relations between astrophysical theory and the theory
of elementary particles can be classified according to the

direction in which information flows from one field to the other and the nature of that information.

First, there are astrophysical-data-set constraints on the development of elementary-particles theories. One example is the limit on the sum of all neutrino masses, set by limits on the present observed deceleration of the cosmic expansion. With increasing accuracy, it will eventually be possible also for solar-neutrino experiments to set limits on off-diagonal elements of the neutrino-mass matrix. These are important results, because some current grand unified gauge theories give the neutrinos small but nonzero masses. Another example is the limit on the total number of neutrino flavors, set by limits on the helium abundance. At present, this provides the most stringent constraint on expectations of new flavors of quarks and leptons.

A very puzzling example is provided by the cosmological constant. Even the crudest astronomical observations set limits on this constant that are incredibly small by elementary-particle standards. It is not known why quantum effects do not produce an effective cosmological constant in excess of these limits.

Conversely, developments in elementary-particle physics have an impact on astrophysical calculations. The best-known recent example is provided by the change in our picture of neutrino interactions arising from the prediction and discovery of neutral currents. This has led to a complete revision of the dynamical calculations of stellar collapse and explosion in Type-II supernovae. Another example is the ongoing improvement in calculations of gross neutron-star properties, including the maximum mass, made possible by improvements in strong-interaction theory. Perhaps the most dramatic example is the recent estimate of the cosmic baryon-entropy ratio that would have been produced by possible baryon nonconserving reactions in the very early Universe.

In addition, astrophysical problems provide a context for calculations that test the significance of elementary-particle theories. For example, it is now believed that phase transitions occurred in the very early Universe, in which the electroweak gauge symmetry and then the strong interaction chiral symmetry were successively broken. These phase transitions have no observational implications that are currently established, but there have been speculations about the development of cosmic domains, in which symmetries are broken in different ways.

A possible consequence of these phase transitions is the development of dynamical perturbations, which will eventually lead to the density perturbations from which structures now present in the Universe form. Since those structures provide evidence on the form and spectra of those perturbations, there is an exciting potential link between the interactions in the very early Universe and the macroscopic structure of the Universe today.

X. EPILOGUE

It is not, after all, surprising that the connections between astrophysics and other sciences should grow richer as astrophysics matures. There are 15 decades between the largest characteristic lengths that can be investigated by astrophysical techniques and the largest (the solar system) accessible to direct measurement by "laboratory" techniques, and nature has provided, in astrophysical environments, extremes of almost all physical parameters far surpassing conditions obtainable in the laboratory. Perhaps the most exciting unrealized area of overlap is with biology (or perhaps even psychology and sociology), but there may well be other unimagined things waiting. The next decade is a very promising one.

6

Astrometry

"We don't know what data astronomers
want in the next twenty years, but we
are sure that they want it with much
greater accuracy."
--- Einar Hertzsprung

I. INTRODUCTION

Astrometry--the determination of positions, motions, and
coordinate systems and the entirety of the products of these
observations--was the first specialty in astronomy. The value
of astrometry for all the fields of astronomy does not have to
be emphasized here; it is clear that the foundation of astronomy
and astrophysics is provided by astrometry. Perhaps not so
clear is the fact that, although considerable progress was made
in the 1970's, the numerical estimates for many fundamental
quantities are much less certain than the average user realizes.
Fortunately, astrometry stands at the brink of a technological
revolution that can provide a tenfold improvement in the
precision of most astrometric observations, with very exciting
implications for all fields of astronomy.

For instance, reducing the standard error of considerable
numbers of trigonometric parallaxes to 1 milliarcsec will
affect all the known distance indicators and even allow the
direct determination of trigonometric parallaxes of a few
Cepheid variable stars. Such precision will also permit direct
determinations of trigonometric parallaxes of the Hyades with
an uncertainty of less than 5 percent. Special ground-based
telescopes will even allow precisions of 0.1 milliarcsec,
leading to the determination of accurate parallaxes out to 1
kpc for small numbers of stars and opening up all stellar types
for accurate luminosity calibrations. The importance of these
advances cannot be overestimated. Work toward this aim should
be generously supported.

The Working Group on Astrometry here discusses the advances that can be made in the 1980's and their impact on the entire field of astronomy. These possibilities and challenges are attracting sophisticated investigators, theoreticians, and particularly instrumentalists to a field in which progress had become somewhat slow. The Working Group believes that these new opportunities and challenges are of such a magnitude that they cannot fail to be given a very high priority commitment of funds during the 1980's, especially since the required expenses are quite small in the context of the enormous gains projected, thus making the investment eminently worthwhile.

In the three sections that follow, Section II lists the major contributions of astrometry, marking the progress of each subarea in the 1970's and the hopes for the 1980's. Section III deals with programmatic opportunities for the 1980's, while Section IV discusses the main instrumental advances that should be achieved, all within the current reach of technology.

II. ASTRONOMICAL CONTRIBUTIONS OF ASTROMETRY

A. Trigonometric Parallaxes

1. Astronomical Impact: Basis of distance scale. Only direct measurement of stellar distances that are the basis for luminosities and the ensuing stellar structure and evolution calculations.

2. Current Precision: 16 milliarcsec, Yale Catalog; 3-5 milliarcsec, present.

3. Progress in 1970's: Precision improved, systematic errors reduced; many faint-star and degenerate-star parallaxes.

4. Hopes for 1980's: 1 milliarcsec ground-based accuracy, 0.1 milliarcsec accuracy from space; trigonometric parallaxes of Cepheids and other now too distant objects should be possible; parallaxes of faint, very red stars may yield the cutoff of the lower end of the main sequence; many more parallaxes for southern hemisphere stars.

B. Catalog Positions and Motions

1. Astronomical Impact: Observations of absolute positions and derivation of absolute motions; extension to fainter magnitudes using differential reference positions and photographic catalogs; basis of fundamental kinematics, cluster parallaxes, solar-system dynamics, contributions to geodesy, etc.

2. Current Precision: ±20 milliarcsec position at epoch (FK4 system); ±1 milliarcsec/year proper motions.

3. <u>Progress in 1970's</u>: W5$_{50}$ AGK3R; SRS observations; initial development of automated transit circle; radar ranging.

4. <u>Hopes for 1980's</u>: Automated devices should show increased observational precision; southern hemisphere transit program; new large-angle measuring techniques, ground-based and spacecraft; FK5; interferometers; atmospheric studies.

C. Cluster Parallaxes

1. <u>Astronomical Impact</u>: First step beyond trigonometric parallaxes in determination of distance scale, used in calibration of other distance indicators.

2. <u>Current Precision</u>: 0.5 milliarcsec or 0.05 magnitude.

3. <u>Progress in 1970's</u>: Systematic errors for Hyades hopefully reduced; distance estimated now 10-15 percent farther.

4. <u>Hopes for 1980's</u>: New parallax and proper-motion studies now under way should be completed; new techniques will allow application to other clusters: Pleiades, Perseus, Praesepe.

D. Stellar Motions

1. <u>Astronomical Impact</u>: Provides cluster internal motions and mass estimates, statistical and secular parallaxes, data for the velocity distribution function, and studies of Galactic kinematics and dynamics.

2. <u>Current Precision</u>: About 20 milliarcsec/year for surveys to 0.01 milliarcsec/year for some studies.

3. <u>Progress in 1970's</u>: Luyten and Giclas surveys completed for high proper motion stars; Lick survey with respect to galaxies 2nd epoch plates; increased precision.

4. <u>Hopes for 1980's</u>: Improved techniques will allow increasingly accurate motions, increasing accuracy of secular parallaxes, etc.; also expect secular acceleration determinations, which, combined with radial-velocity measurements, allow direct parallax measurements.

E. Binary Stars

1. <u>Astronomical Impact</u>: Determination of stellar masses; basis for stellar structure and evolution studies.

2. <u>Current Precision</u>: 40 milliarcsec visual observations; ±20 percent for masses calculated from reliable orbits; 3-milliarcsec speckle interferometry observations.

3. <u>Progress in 1970's</u>: Speckle interferometry became practical.

4. Hopes for 1980's: Improvement of speckle interferometry; single- and double-aperture interferometers
developed to aid/complement visual observations.

F. Stellar Diameters

1. Astronomical Impact: Necessary for the study of
stellar atmospheres, structure, and evolution.
2. Current Precision: 1 milliarcsec.
3. Progress in 1970's: Majority of measurements made in
this decade, by occultations and interferometry.
4. Hopes for 1980's: Improvement in interferometric techniques should allow many more observations.

G. Astrometric Binaries

1. Astronomical Impact: Discovery of low-mass stellar
companions; most promising method to search for extrasolar
planetary systems.
2. Current Precision: 7 milliarcsec smallest detected.
3. Progress in 1970's: Discovery rate increased mainly
due to increased accuracy of measuring and reduction techniques.
4. Hopes for 1980's: Improvements expected in instrumentation should allow the determination of the proportion of stars
with stellar companions and possibly the discovery of other
planetary systems.

H. Solar-System Objects

1. Astronomical Impact: Testing of gravitational theories;
discovery of unknown objects; determination of astronomical
constants; evolution of solar system; use for space navigation.
2. Current Precision: 0.3-arcsec transit circle;
70-milliarcsec photographic; 250-m radar ranging; 15-cm laser
ranging; 100-m orbiter ranging; 0.1-0.01 sec of time in
occultations.
3. Progress in 1970's: Introduction of laser and
spacecraft ranging techniques; discovery of satellites and
rings around planets; new generation of planet and satellite
ephemerides.
4. Hopes for 1980's: Continued systematic observations
using improved techniques and with improved accuracy; observations from Space Telescope; increased knowledge of astronomical constants, gravitational theories, evolution and
structure of solar system, and objects in solar system.

III. PROGRAMMATIC OPPORTUNITIES FOR THE 1980'S

A. Parallaxes

The improvement in the precision of trigonometric parallax
measurements from the early 1960's through the 1970's amounts
to a factor of about 4. Much of this improvement has been
achieved with conventional long-focus astrometric refractors
that have been in use for decades. The higher precision has
been due to the introduction of automatic measuring machines,
automatic guiding, high signal-to-noise emulsions, and more
rigorous computer modeling of the telescopic field, often
coupled with the use of stellar image profiles. The use of the
U.S. Naval Observatory's 1.55-m reflector has introduced a
further reduction in error among parallaxes; this telescope was
designed to eliminate many of the problems inherent in the
older refractors and has provided not only a large number of
highly accurate parallaxes but also a benchmark against which
new parallaxes determined with the old refractors can be compared. The 4-m reflector at Kitt Peak National Observatory has
been used recently in a pilot program to determine the most
precise unit-weight parallaxes measured to date. Finally, most
existing parallax programs, internationally as well as within
the United States, are now becoming coordinated through the use
of a list of parallax standard stars covering the entire sky.
These efforts will calibrate all telescopes onto a common system with a known zero point for the first time.
   We are now at the threshold of being able to determine
parallaxes with an accuracy of ± 1 mas (throughout the rest of
this chapter 1 mas = 1 milliarcsecond = 0.001 arcsecond) from
only a few dozen plates, using finer-grained emulsions and
sophisticated image analysis. It should be noted, however,
that in addition to parallaxes with ± 1 mas precision for some
of the most important stars, a much larger number with the
currently attainable precision of about ± 4 mas will be obtained
in the 1980's. There are several reasons for continuing
long-term support of the present parallax efforts as well as
encouraging the development of new equipment and procedures
that are expected to offer the greater precision. First, the
equipment is in place and operating, and the programs have
already amassed large numbers of photographic plates. Second,
it is not likely that parallaxes with an accuracy of 1 mas can
be determined for more than a few stars of any one kind even by
the end of the decade. Parallaxes of large numbers of stars
must be measured to answer many questions, and these will only
be available from continuing programs.
   The list of stars that have had a trigonometric parallax
measurement reflects the programs at the various observatories
that make these measurements. Because the accuracy of a trigonometric parallax is a function of the equipment and measure-

ment procedure at a particular observatory, there is a rough correlation between the type of star and the precision of its parallax. Generally speaking, the stars with measured trigonometric parallaxes (published prior to 1980) that will appear in the new Yale Parallax Catalog fall into three groups:

1. Almost all stars brighter than magnitude 4 and a substantial number with magnitudes between 4 and 6--typical accuracy ± 10 to ± 15 mas;
2. A substantial number of fainter stars of high proper motion, largely lower-main-sequence stars--typical accuracy ± 15 to ± 20 mas;
3. A smaller number of intrinsically faint stars, largely K and M main-sequence stars and white dwarfs--typical accuracy ± 4 to ± 8 mas.

It should be noted here that the old Yale Parallax Catalog has an average standard error of 16 mas, with possibly large additional systematic errors. Yet this catalog is used regularly by astronomers who are interested in individual distances and/or luminosities of a star or of a whole class of stars. The new Yale Catalog, now in preparation, will still leave large gaps in the completeness of the parallax data, and many parallaxes of bright stars can be vastly improved using the currently attainable 4-mas accuracy.

Since current programs concentrate on the white-dwarf and lower-main-sequence stars, the data being accumulated will bear on the lower part of the H-R diagram. The low-luminosity cut-off of the main sequence, fine structure in the lower main sequence arising from differing chemical composition, the relation of this fine structure to the age of the Galaxy, and fine structure in the lower part of the mass-luminosity diagram are among the problems that can be discussed using the results of current programs. The parallax precision attainable at present and the improvements anticipated for the near future promise parallaxes that will provide detailed information about stars in other regions of the H-R diagram. The new questions that will be addressable include the following:

1. Fine structure in the central main sequence. It has been suggested that variations in helium abundance may produce detectable fine structure along the main sequence.
2. Metal abundance versus age. This question cannot be adequately approached until separate evolutionary sequences can be traced for stars with different chemical composition. Accurate parallaxes are required.
3. Luminosity classification. Good parallaxes are important for luminosity classification over a considerable part of the H-R diagram, and luminosity classification is central to problems such as the zero-age main sequence,

evolutionary tracks, and the Galactic and cosmic distance scales. A substantial number of giant-branch stars are close enough for good parallaxes with current precision.

4. Stellar masses. Only a few subgiant stars and one giantstar system have accurate masses. Lists of binary stars believed to be "nearby" (based on dynamical parallaxes) include not only several red giants but also Ap and Am stars and main-sequence stars up to middle B. Masses of single stars, although not fundamental in nature, can be determined from detailed spectroscopic analyses. Such analyses yield the surface gravity, which, coupled with the radius, gives the mass. The greatest uncertainty in the radius now comes from the parallaxes.

5. Age of the Galaxy. The age of the halo population of our Galaxy is obtained by fitting theoretical evolutionary tracks to color-magnitude diagrams of globular clusters; the latter requires accurate parallaxes of subdwarfs.

In addition to the above (admittedly incomplete) listings, it should be noted that as ± 1 mas accuracy is approached, a few rare but extremely interesting stars will come within range of parallax measurements, including some RR Lyrae and Cepheid variable stars.

One should bear in mind that the correction from relative to absolute parallax, and its uncertainty, are currently also of the order of 1 mas. Hence, care will have to be taken to determine this correction more precisely. This mainly requires good spectrophotometry of the reference stars. If we assume a correction from relative to absolute of 1 mas for 16th-magnitude reference stars, an error of the spectrophotometric parallax of 0.5 magnitude, and 10 reference stars, we find an error of 0.1 mas for the correction. Therefore, there will arise an extensive need for excellent photometry connected with parallax measurements, but absolute parallaxes with 1-mas accuracy can easily be attained. Other possible methods are deconvolution of the parallax distribution, measuring the parallaxes of a large number of reference stars as well as of the target star, and the unlikely presence of a quasar in the parallax field.

In conclusion, it should be emphasized that the maintenance of support for existing parallax programs and the need further to develop and refine parallax methods of even higher precision are mutually supportive; both should receive high priority. Since most parallax programs are conducted at colleges and universities, a vigorous endorsement is especially called for in view of their well-known reluctance to encourage long-term research, which is often without immediate or sensational results.

B. Stellar Diameters

Recent developments in stellar interferometers have made the regular measurement of stellar diameters and such extended objects as circumstellar dust shells a probable activity for the 1980's. The development and use of a Michelson interferometer on the 5-m Hale Telescope, together with the development of infrared and optical two-telescope interferometers, indicates that resolutions of 1 mas or better may be regularly obtained during the 1980's. Possible programs include the following:

    1. Emergent fluxes and effective temperatures for single stars;
    2. Stellar radii;
    3. Limb darkening;
    4. Stellar rotation;
    5. Extended atmospheres of early-type stars;
    6. Circumstellar dust clouds;
    7. Emission-line stars;
    8. Interstellar extinction;
    9. Cepheid variables; and
   10. Galactic nuclei and quasars.

This list is not intended to suggest an order of importance, nor to be exhaustive, but rather to indicate the wide range and potential of stellar interferometry.

C. Double Stars

The number of known visual double stars exceeds 70,000 and continues to grow. The number of observers, however, continues to decline; as a result, thousands of pairs showing orbital motion are necessarily neglected. At the same time, prospective improvements in parallax techniques during the 1980's afford the opportunity of increasing the number of accurately determined stellar masses from about 50 to some 500, provided that double-star observations are continued. (Note that the percentage error in the masses is three times the error in the parallaxes.) Intrinsically brighter stars will thus have good mass determinations for the first time, and the entire empirical mass-luminosity relation will be both broadened and strengthened. Fine structure of the M-L relation arising from differences in age, chemical composition, and possibly other parameters may become visible for the first time, with obvious implications for studies of stellar structure and evolution.
The application of interferometric techniques holds bright promise for the resolution of close binaries in the range 0.01-2.0 arcsec and in addition appears to be more accurate

than the visual measures made heretofore. A considerable
number of spectroscopic binaries can already be resolved by
existing techniques. Moreover, the development of rapid and
much more accurate means of radial-velocity determination will
considerably increase the number of suitable systems for which
the combination of precise interferometric and spectroscopic
observations will permit the determination of absolute parallaxes and masses. Already over a dozen reliable orbital
parallaxes have been measured.

Various types of interferometers are either in use or under
active development at present. All of these have the potential
to contribute to binary-star studies. By the end of the decade,
long-baseline optical interferometers may be operating in the
milliarcsecond range or below. At present, the speckle technique is yielding extremely valuable results in the range
0.04-1.0 arcsec. It appears likely that a relatively simple
and compact speckle system can be constructed for use on
telescopes of moderate as well as large aperture. In view of
the high rate of data acquisition of this technique, a significant increase in positional data for at least the brighter
binary stars is possible. Development of such a system should
therefore have high priority. Toward the end of the decade,
long-baseline systems may finally eliminate the classical gap
between the visual and spectroscopic binary; such systems
should be supported as a follow-on to the more modest capabilities of the single-aperture system.

The benefits of improved precision and higher acquisition
rate are not merely a linear function of these data because of
the long periods of most visual binaries, which does not allow
accurate determination of the major semiaxis. The improved
precision and data density that we expect for the 1980's,
however, will permit the determination of orbits (and thus
masses) for many more long-period pairs. Finally, it must be
emphasized that because of the time constraints imposed by
nature, programs must be planned and sufficient telescope time
allocated on a more or less continuous basis to define the
orbital motion reliably in each individual case.

D. Extrasolar Planets

The search for (and, it is to be hoped, the study of)
extrasolar planetary systems is most easily carried out by
astrometric techniques. The resolution of a planet from its
parent star is considered nearly impossible for all but the
very easiest cases, and the radial-velocity variations (which
amount to 10 m/sec or less) require extremely sophisticated
equipment for their measurement. Therefore, astrometry remains
the most sensitive technique for finding extrasolar planets.
The near-circular orbits assumed for planetary systems induce

perturbations in the motion of the primary stars that are
almost equally detectable from any orientation in space.
Furthermore, these deflections from linear proper motion are
relatively large; there are 15 stars observable from northerly
latitudes whose perturbation by a Jupiter-like planet would be
greater than or equal to 4.5 mas.

The mass of the smallest detectable planet is almost a
linear function of the precision of the astrometric instrument. An Earth-like planet would induce a perturbation of
approximately 3 μarcsec in the motion of a solar-type star
seen at a distance of 1 parsec. This effect may possibly be
detectable with a specialized astrometric telescope in space.

The limit to detectable masses probably comes about as a
result of the apparent photocentric shifts caused by starspots. However, the effects of the largest starspots can be
recognized and modeled, yielding rotation rates and axis
positions for the star and the spot's latitude--extremely
important astrophysical quantities. The limit to detectable
planets will be encountered when large numbers of starspots
cause an apparent random noise, which, for the Sun, may at
times be large enough to conceal the pull of Mercury.

Planetary detection is one of the most exciting new fields
of modern astronomy. Each study will produce accurate
astrometric parameters for at least one target object (two for
binaries) and a score of reference stars. Studies of stars
with unseen stellar companions (neutron stars, black holes)
fall in the same category. Studies of such objects with interferometers or specialized astrometric telescopes down to 0.1
mas will dramatically alter this field of astronomical research.

E. Proper Motions

A knowledge of the proper motions of large numbers of stars has
a direct and fundamental impact on many basic areas of
astronomy. Three such areas are stellar kinematics, stellar
dynamics, and clusters and associations. Each of these, in
turn, can be broken down into a number of subareas or problems
as follows:

1. Stellar Kinematics
   a. Local space motions of different types of stars
   b. Velocity gradients
   c. Solar motion
   d. Galactic rotation (Oort's constants)
   e. Statistical and secular parallaxes
   f. Comparison with kinematics of gas and dust
   g. Determination of precession
2. Stellar Dynamics
   a. Tidal effects--age/size/density of clusters and
      associations

       b.   Spiral structure and its evolution
       c.   Orbits of halo and disk objects
       d.   Galactic gravitational potential

3. <u>Clusters and Associations</u>
       a.   Membership
       b.   Distance and calibration of the magnitude scale
       c.   Expansion/contraction, internal kinematics
       d.   Relation between kinematics and age/chemical
            composition

Improved observational techniques of the next decade (e.g., photoelectric transit circles, space and interferometric astrometry) will ultimately lead to proper motions of substantially better accuracy and precision than any currently available. Furthermore, the extension of a proper-motion system with improved accuracy to a fainter magnitude range will lead to the expansion of the sphere around the Sun in which an unbiased kinematic sample may be obtained.

At this time, limitations of accuracy and precision of the motions themselves, together with the small size and highly biased nature of the sample for which these are available, are forcing on us a highly simplified and certainly unrealistic model (Galactic rotation, ellipsoidal distribution) for Galactic kinematics, with the concomitant paucity and uncertainty of those data against which the various theories of Galactic dynamics are tested. It will require a massive effort for the improvements of all aspects (quantity, precision, accuracy, diversity, uniformity of sky coverage) of proper motion research before this situation will be more favorable.

An immediate result of eventually available proper motions will be improved kinematic parameters, including more accurate secular and statistical parallaxes for the disk stars and the halo stars on which studies of their luminosities and kinematics depend. These stars include types crucial to the determination of the distance scale and to our understanding of the details of stellar evolution, such as the RR Lyrae and CH stars. An unbiased sample of low-luminosity main-sequence stars will help to determine the luminosity function in this range. The Lick Proper Motion Survey with respect to faint galaxies in the northern hemisphere and its southern counterpart organized by Yale University at El Leoncito, Argentina, are designed to provide some of the above listed basic data. For example, they might improve the parameters describing the velocity field of the centroids (in simplified form these are Oort's rotation constants of the Galaxy) and the determination of an inertial frame of reference (precession, motion of the equinox). The second phase of the Lick Survey will continue through the 1980's. It is at this time planned to have the third phase of the Lick Survey begin in the 1990's when it will be possible to utilize the first-epoch yellow plates taken in

the 1970's and the blue plates taken around 1950. The expected
increase in accuracy, precision, and quantity of available
proper motions makes this a program of very high priority,
especially in view of the fact that the contemplated tie-in to
extragalactic (and therefore presumbly without detectable
motion) objects should constitute a large step in the direction
of substantially improved accuracy by shedding the traditional
dependence on the Earth's kinematics. Yale, for the purpose of
its Southern Survey, was fortunate to have had both lenses
installed in time for the first-epoch photography, so that its
second-epoch photography (planned for the late 1980's) will use
both blue and yellow plates. The measurement and analysis of
these data will extend at least well into the 1990's.

More precise proper motions will allow one to investigate
more closely the internal motions in an increasing number of
Galactic and globular clusters and, among other things, to
determine their virial masses. These data are of great value
to theoreticians who need measurements of the degree of isotropy and the radial distribution of the internal motions to
discriminate between the stellar dynamical models of star clusters. In the next 10 years we should have available accurate
measurements of the internal motions of several globular and
open clusters of a variety of types to aid in the definition of
suitable dynamical models and in the determination of virial
masses and kinematical distances to these clusters.

Other study areas of the future that depend directly on
accurate proper motions include investigations concerned with
more adequate modeling of the velocity field of the velocity
centroids and the establishment of the dependence of the
parameters in the velocity-distribution function on location
within the Galaxy, for example, possible asymmetries of stellar
motion perpendicular to the galactic plane and the behavior of
the Galactic potential in this direction. It should also
become possible to discover the relationship between kinematic
properties and physical characteristics, such as chemical composition in various parts of the Galaxy, to test the reality of
the missing-mass hypothesis, to improve the determination of
the dynamical local standard of rest, and to derive a better
understanding of star formation and spiral structure from
improved analysis of Gould Belt dynamics.

Finally, increased accuracy and fainter limiting magnitudes
will permit proper motions to be used for membership segregation of intrinsically faint stars in clusters at increasing
distances and over a wide age range. This will allow the
determination of the faint end of the luminosity function in
these clusters, the location of the lower end of the main
sequence in an H-R diagram, and a search for white dwarfs, as
well as determining the membership status of individual stars
of interest.

F. Catalogs of Relative Star Positions

The "fundamental system," which is the best approximation to an inertial system currently available, consists of about 2000 positions and proper motions, which were compiled to form the FK4; these are the fruit of the labors of several generations of meridian astronomers who have worked with the world's transit circles for almost the last 200 years. These observations were largely made absolutely--that is, in such a way that the establishment of the coordinate system to which they are referred is an integral part of the measuring process.

Within the last 25 years, a secondary system of reference stars based on transit-circle observations (and referred to the fundamental system mentioned above) was established, consisting of about one star per square degree. These stars served as a reference system for a tertiary set of photographic catalogs, e.g., AGK3 and Cape Photographic Catalogue.

Finally, a fourth, photographic, tier is provided by the astrographic catalog, the first huge international astronomical enterprise, which involved about 20 observatories, with an average of about 50 stars per square degree to about 13th visual magnitude with a total of about 2.5 million stars. This material provides the positions of stars for an average epoch of about 1890 with a standard error of 0.25 arcsec or better. Combination of these data with modern data can provide proper motions for all the stars in the original catalog with a standard error of no more than 4 mas/yr. In view of the importance of the availability of a large number of precise and accurate proper motions, this project should receive high priority.

The strengthening--that is, improving the accuracy, precision, and density--of the stellar reference frame is a never-ending process, and several projects in several countries are contemplated in pursuit of this goal. The most ambitious is a 10-year program now under way at the U.S. Naval Observatory: an astrographic survey of the entire sky down to 13.5th magnitude in two colors, yellow and blue, which will lead to a positional catalog of 10 million tertiary standard stars, free of the magnitude and color errors known to affect earlier work. Provisional results indicate that positional accuracy between ± 0.1-0.2 arcsec is being achieved.

G. Fundamental Astrometry

The fundamental reference systems and the system of radial velocities are the foundation upon which essentially all knowledge of the kinematics of the Universe rests, either directly or indirectly. In addition, there rests upon this foundation the calibration of a vast inductive astrophysical extrapolation.

The fundamental system, in turn, rests exclusively upon so-called "classical" meridian astrometric observations. During the last one half to three quarters of a century, the commonly recognized astronomical community of the United States, i.e., various universities and associated or independent observatories, together with their funding sources, have contributed nothing to the establishment or improvement of the fundamental system. Hence, when one speaks of fundamental astrometry, one is speaking of the astrometric efforts of the U.S. Naval Observatory. The contribution of the United States to fundamental astrometry is that which is gleaned as a by-product of satisfying the requirements of the Navy. The principal reason for the lack of research in fundamental astrometry in other groups has obviously been the long-term nature of the observational programs, which do not lend themselves easily to the publication requirements imposed by teaching institutions and funding agencies. Fortunately, the situation is changing.

During the past decade several new techniques have emerged that can contribute to fundamental astrometry. These techniques range from the application of new technology to classical methods, on the one hand, to the new observing techniques exemplified by interferometric methods on the other. Observations from space have been planned or proposed, but as currently envisaged they will not be absolute in nature but provide only internally self-consistent systems of positions and proper motions. In an interactive mode, space and ground-based systems may become mutually supportive.

One of the most exciting developments of the 1970's was the development of radio astrometry, to the point where absolute declinations and absolute right ascension differences of objects with sufficiently strong radio brightness (quasars) can be determined more accurately than the corresponding data for optical objects (stars). During the 1980's, developments aimed at increasing the sensitivity of the radio-positioning methods, as well as the placing of an active satellite into a heliocentric orbit for the purpose of determining the vernal equinox (and thus essentially an inertial system) by radio means, deserve strong support. It should be noted, however, that such radio methods will essentially be limited to a small number of extragalactic objects and will by no means remove the burden of fundamental optical astrometry needed in studies of our stellar environment.

The coming decade should see much activity and progress in fundamental astrometry, particularly in the following broad areas: improvements to current methods, interferometry, astrometric investigations, and observations from space. These are discussed in the next section of this chapter.

In principle, stellar proper motions can be referred to an inertial reference frame determined by galaxies and quasars. In addition, the quasar reference frame determined by radio

astronomical methods is a system of fundamental positions with
potentially a much greater accuracy than the optical system.
Unfortunately, the finite size of galaxies and the paucity and
faintness of quasars has so far prevented the establishment of
a fundamental proper motion system referred to extragalactic
objects. The major effort by the Lick Observatory, as well as
other efforts in this direction, should be strongly supported
in the 1980's.

At the same time, transit-type observations must be continued to provide input into the empirical establishment of the
kinematics of the Earth, which cannot be obtained from an
inertial system of proper motions established only by reference
to extragalactic objects. This is the more important since the
improved precision of transit-type observations (transit
circle, PZT) has opened the door to fuller empirical monitoring
of the deviations of the Earth's rotation from rigid-body
behavior. These observations will have continued undiminished
value for astronomy as long as there remain the currently unexplained differences in the precessional parameters (including
motion of the equinox) obtained from observations by transit
circles and the extragalactic reference frame, respectively.
Even so, certain branches of geophysics and geodesy depend on
transit-circle observations for their only source of certain
types of data, and fundamental astronomers must continue to
fulfill their historical obligations to geodesy and geophysics.

H. Solar-System Astrometry

Continuous series of accurate observations are required for
solar-system bodies in order to determine the orbital motion of
the bodies. These observations must be referred to an inertial
system, because only in such a system can the results from the
properly integrated equations of motion legitimately be compared with observations; such observations should be made by a
variety of means, including radar and laser ranging, photography, occultations, transit circles, and spacecraft, by
several observatories on a regular basis.

The wide range of magnitudes and sizes of solar-system
bodies presents special problems: the different observations
must be placed on the same coordinate system, which requires
special efforts for faint objects and nonoptical observations. In many cases, the accuracy of the reference-star
position is the limiting accuracy for the observations.

Special efforts are required to obtain observations of
faint satellites, satellites close the primary, and rings.
There should also be concerted efforts to search for undetected
satellites, asteroids, and comets. Unexplained perturbations
in the orbital parameters of the outer planets make the
presence of additional planets a likely occurrence; they will

probably be quite faint, however, and will require good predictions or thorough searches.

The instrumentation discussed in Section IV for both
fundamental and relative position measurements, ground- and
space-based, will be used for many of the solar-system observations in the radio, infrared, and optical wavelength regions.
We wish to highlight two additional methods here that will have
a considerable impact on the field in the 1980's.

1. Direct Distance Determinations

Until the 1960's, celestial mechanics had only angular positions
of objects to use as fundamental data for orbital theories.
Distances had to be inferred indirectly. Even the fundamental
scale of the solar system was known to only three significant
digits, and, as it turned out, there were important systematic
errors in that determination.

The building of large radar "dishes" in the 1960's fundamentally changed this situation. It became possible to determine the distances to the planets with far greater accuracies
than before; measurements to a fraction of a kilometer became
routinely available. This increased accuracy made possible far
more precise determinations of planetary orbital parameters
than ever before, as well as improved masses of the planets.
Tracking data from planetary probes provided additional precise
determinations of planetary masses. So accurate were these new
data that the effects of the General Theory of Relativity had
to be included in the mathematical models used to reduce them.

The landing of Apollo 11 on the Moon in 1969 led to another
era in the field of direct-distance measurements. The retroreflectors left by Apollo astronauts and unmanned Soviet lunar
landers have been used in a continuing laser-ranging program
ever since they were emplaced. So precisely can the returning
light be timed that the distance to the Moon can now be determined routinely to 10-15 cm. The implications of this in many
areas have been profound. The accuracy of these data challenges
the precision available from present-day orbital theories of
the Moon. They have made possible new tests confirming
Einstein's General Theory of Relativity, and independent data
of great precision have been obtained on the rotation of the
Earth and the librations of the Moon. These data will also be
important in the next decade as part of efforts to determine
whether, indeed, the constant of gravity is changing with time,
a situation that would have profound astrophysical as well as
cosmological consequences. Closer to home, lunar laser ranging
is providing fundamental geophysical data, which, in the next
decade, will provide direct measurement of continental drift
and better understanding of earthquakes and their causes.

It is the nature of observations of the motions of celestial bodies, such as these, that they need to be carried out on

a sustained and systematic basis. Valuable as the data already obtained by these techniques have been, unless they are followed up, their ultimate value will be greatly diminished.

Lunar laser-ranging equipment has come a long way, and the retroreflectors on the Moon are likely to remain effective for many decades. Improvements in lasers and control electronics have made possible much more compact equipment utilizing smaller dedicated telescopes. Geophysical as well as astronomical considerations indicate the need for a well-dispersed network of lunar-ranging stations, including mobile ones, to exploit this technique maximally.

2. Occultation Observations

In the 1970's, a marriage between high-speed electronics and microcomputers made high-speed occultation observations possible in the visible and infrared part of the spectrum. Three main results of this work have been obtained: precise observations of the motion of the Moon, observations of multiple stars and stellar diameters, and new information on the nature of bodies in the solar system. The observations of lunar motion have been important in research suggesting that the constant of gravity may be variable; this fundamentally important question has not been resolved, however, and more occultation data are needed. Very precise observations of close multiple stars have been made using occultation techniques. These data cover a range of separations that have been very difficult to obtain heretofore, a range where statistics on stellar duplicity are poorest. Although only about 10 percent of the area of the sky is accessible to such observations, the question of the frequency of duplicity is of fundamental importance in understanding the formation of binaries and in helping to resolve the question of the existence of extrasolar planetary systems. Observations of occultations by minor planets have revealed the surprising possibility of the existence of minor planets with one or more satellites, as well as providing direct information on the shapes of minor planets. Occultation observations of Uranus revealed first the existence, and then the surprising dynamical features, of its rings.

New techniques will make the prediction of occultations of stars by planets and minor planets a routine operation. Support is needed in the 1980's for the implementation of these techniques and for doing the actual predictions.

Instrumentation for high-speed occultation observations is cheap and is getting cheaper. There are many small telescopes both in the United States and around the world that could make such observations routinely if they were properly equipped and funded. Moreover, there is a great deal of interest among astronomers and physicists at institutions with small telescopes in performing observations of this kind, which are both within their reach and of scientific value. A network of occultation observation stations, associated with existing small observatories, should be supported.

This program would provide data for the improvement of the lunar theory and for the detection of a possible variation of the constant of gravity; it would also provide routine observations of multiple stars, both for the detection of duplicity and for the determination of orbits (which in turn will help improve our knowledge of the mass-luminosity relation). It should be noted that it takes two stations, observing the same occultation event, to measure both separation and position of a pair of stars.

I. Earth Motions

During the 1980's, activity in the field of Earth motions will be concerned with the continuation of regular observations of the orientation of the Earth in an inertial reference frame, improvements in present observational techniques, development of new observational techniques, and the analysis of the observations in relation to geophysical phenomena. Classical observations describing the motion of the Earth in the fundamental optical reference frame will be continued internationally (i.e., photographic zenith tube, Danjon astrolabe, visual zenith telescope). Present techniques, which show an approximate accuracy of ± 9 mas for 5-day means of the angles describing the Earth's motion (derived by averaging data from about 60 observatories), may be improved with continued research into the effects of refraction in systematic errors. Available technology will allow the description of refraction anomalies, which can be used to improve these data. Doppler observations of military navigational satellites will continue to provide independent estimates of the position of the pole. Currently achievable accuracies of ± 6 mas in the polar coordinates will permit the evaluation of the systematic errors in the classical data. The possible replacement of these satellites with the Global Positioning Satellites and the resulting loss in accuracy for the polar coordinates suggests, however, the consideration of the possible need for the establishment of some satellites dedicated to the Doppler determination of polar coordinates. Laser ranging to the Moon and to artificial Earth satellites will be further developed in the future. Preliminary results from these techniques indicate that the 1980's might well see accuracies of approximately ± 2 mas in describing the orientation of the Earth in the dynamically defined coordinate system. Full use of these techniques, however, requires dedicated day-to-day operation in order fully to evaluate possible systematic differences. The use of both intermediate and very-long-baseline interferometry (VLBI) is currently being developed. Once the influence of the Earth's atmosphere can be corrected for, accuracies as high as ±2 mas might be possible for the determination of the orientation of the Earth within a reference frame defined by the positions of radio sources. Again, full utilization of this technique will depend on the existence of a dedicated daily operation.

Continued observations, together with anticipated improvements, using the classical instrumentation along with the full development of the new techniques will provide the most accurate set of data available to describe the orientation of the Earth in various reference frames. Analysis of these data will lead to a more complete evaluation of the motions of the basic astronomical reference frames, including precession, nutation, and polar motion. In addition, the validation of geophysical theories and nonrigid Earth models will be possible with improved accuracy in the Earth-orientation parameters.

IV. INSTRUMENTATION

A. On-Line Measuring Devices

Five characteristics that should be embodied in an ideal astrometric detector are the following:

* High spatial resolution;
* High quantum efficiency;
* Freedom from sensible systematic error;
* High precision; and
* Ability to provide concurrent records of a sufficient number of reference stars to allow optimum use of the optical system.

Astrometry's standard "detector" is the photographic plate. This detector has unmatched storage and archival properties, but recent studies indicate that it may limit the ultimate accuracy achievable by astrometric methods. The deleterious effects on the accuracy stem not only from the intrinsic properties of the photographic plate (e.g., relatively low quantum efficiency, emulsion shifts, and nonlinearity) but also from the need for an additional manipulation involving measuring machines. A measure of the accuracy attainable on a high-quality telescope, such as the U.S. Naval Observatory's 1.55-m telescope at Flagstaff, Arizona, is that a good-quality exposure yields positions accurate to 20 mas (s.e.). This error includes contributions from measuring errors and from emulsion shifts. The use of finer-grained emulsions (e.g., III aJ) and improved measuring machines increases the attainable accuracy by a factor of 2 to 3 but seems unlikely to yield gains of a factor of 10 to 30 that appear to be realizable with photoelectric detectors. We do not suggest that further attempts to improve the accuracy of photographic astrometry be ignored; che photographic plate will still have a role to play. Rather, we suggest that an increased emphasis be placed on the development and testing of new detectors that could provide a quantum-leap increase in accuracy.

Recent advances in photoelectric sensors and inexpensive computers have spawned several proposals for new high-precision focal-plane detectors. These fall into two groups--those that measure the position of each reference star and those that sense the average position of a large number of faint reference stars. Detectors in the latter class must be mounted on specially designed telescopes that would be free of sensible systematic errors.

There are two subclasses to the first group of focal-plane detectors--those employing photomultiplier technology and high-precision ruling, and those employing charge-coupled devices (CCD) and charge-injection devices (CID). The high-precision ruling type detectors can generally cover larger fields and have a higher potential spatial precision than the CCD's. They are also self-calibrating. The CCD's and CID's are currently confined to small fields. While the approaches of these two subclasses are more costly than the other group (those finding a single average photocenter for all the background stars), they retain all the dimensional information available for each star and for the optical system. It is therefore possible to model actively and to remove the sources of error introduced by changes in the optical system and the atmosphere.

The devices proposed to date vary considerably in detail and to some extent in application. CCD arrays are ideal for very small fields and speckle interferometry. Image dissector tubes or moving fine-lined rulings combined with high-speed photometry can be used to measure the fringe visibility in Michelson stellar interferometry. Both approaches offer eventual sub-mas precisions in the measurement of double-star separations and can also yield photometry and colors for the components.

The wider-field detectors now being developed use either a mask or a Ronchi ruling as a spatial metric. Those detectors employing Ronchi rulings generally require a separate photometric system for each star and are thus limited to a small number of reference stars, but they have the advantage that the information for each star is retained. Such a prototype detector has been tested and has achieved precisions of 4 mas in 8 min; this system promises to achieve 1 mas/hour precision in its final form, a limitation set by atmospheric turbulence. Unfortunately, existing telescopes will probably not be stable to these precisions and will limit astrometric observations

with them to precisions of a few mas/hours. Thus, new specialized telescopes will have to be designed to make full use of
expected detector accuracies, as discussed below.

In view of the information that will be made available to
astronomy by the quantum leap in precision offered by the new
on-line focal plane detectors, the Working Group favors the
following:

1. The development of several detector designs,
2. The deployment of these new detectors on existing telescopes, and
3. Studies to determine how the characteristics of these
detectors will influence the design of new ground-based and
space-based astrometric instruments.

Interferometric instruments attached to existing telescopes
constitute an entirely different class of on-line devices. The
Michelson interferometer in its various forms, as well as the
speckle interferometer, have come into their own during the
1970's as powerful means of reaching the theoretical telescope
resolution in spite of atmospheric effects. For the determination of stellar diameters and in double-star work, these
instruments have already proven their worth. In both cases,
the next step forward will come, as in the case of one group of
focal-plane detectors, with the advent of large-array detectors. We can look forward to direct imaging capabilities of
small fields in the first half of the 1980's, combined with a
much larger dynamic range than is generally available photo-
graphically.

The Working Group believes that these detector developments
should be vigorously supported. They promise a major improvement in accuracy and in fact will open up an entirely new
field, imaging at the 0.1-arcsec level and beyond, with a minor
expenditure of funds.

B. Telescopes

The greatest part of narrow-field astrometric work has been
done with old refractors, few of which were designed for that
purpose. Some astrometry has also been done with Cassegrain
and Ritchey-Chrétien reflectors, but they too have not usually
been designed for astrometry. Only in recent years have
reflectors been built specifically for astrometric studies; the
most active of these is the 1.55-m reflector of the U.S. Naval
Observatory (USNO) at Flagstaff, Arizona.

An ideal astrometric telescope should have the following:

1. Symmetrical images (i.e., no coma);
2. Complete uniformity of images over the field (i.e., no
axis or no field center);

3. No drifts due to temperature, flexure, or aging effects and a means for checking these stabilities;

4. No chromatic aberrations, even when the atmosphere is considered (i.e., the telescope-atmosphere system should be achromatic); and

5. All optical components located at the system's pupil. Existing refractors have some residual coma, large chromatic aberration, and flexure of the long tube (at f/15 to f/20). Flexure and temperature effects within the lens have also been noted. Either slow unchecked aging or abrupt changes (after dismounting and remounting the lens, for instance) are serious defects for very long-term programs.

The USNO astrometric reflector has relatively large but highly stable coma. However, the effect of coma can be minimized with a narrow field of view and with linear detectors. From recent results, it appears that this reflector is free of systematic errors down to the present accuracy limit for photographic results (e.g., a mean 1-mas error in parallax studies). It is also clear that one cannot speak of telescope errors independently of plate-reduction techniques. Modern error-modeling procedures are able to accommodate field effects that used to be disastrous; hence, the design details of a new astrometric reflector are not yet certain.

Unfortunately, only one such telescope is available full-time for astrometric work. Using some of the on-line measuring devices described above, both the magnitude limit and the accuracy could be improved considerably. Even more importantly, however, the fact that no such capability exists in the southern hemisphere prevents the investigation of objects in almost half the sky and particularly limits the study of objects concentrated toward the Galactic center and the richest part of the Milky Way.

The case for placing the next astrometric reflector in the southern hemisphere is clear-cut. First there is the problem of statistical completeness. Such questions as a possible asymmetry of the stellar luminosity function, the out-of-plane motion of the Sun, the complete local structure of the Galaxy, and galactic kinematics beyond the simple rotation models can only be answered with comprehensive full-sky astrometry. The richness of the southern Milky Way, in contrast to that visible from the northern hemisphere, allows much more intensive studies of both open and globular star clusters. Determination of parallaxes of southern stars of special astrophysical interest (as contrasted to a general parallax survey) will require large amounts of astrometric observing time. Parallaxes of stars in the direction of the Galactic center suffer from the large zenith angles at which they have to be observed. And, even in the parallax case, we might well run into statistical completeness problems: Is it really evident that the stellar population in the celestial northern hemisphere is representative of the whole?

This Working Group therefore believes that high priority should be given to a design study and subsequent construction of an astrometric reflector in the 1.5- to 2.5-m class in the southern hemisphere, equipped with state-of-the-art measuring devices.

While existing large Ritchey-Chrétien telescopes, both in the northern and southern hemispheres, lend themselves to narrow-field astrometry with surprising accuracy, such telescopes are so heavily oversubscribed that astrometric work occupies only a small fraction of the observing time. In particular, the long-term nature of astrometric projects is a great drawback for the appropriate allocation of observing time. The Working Group urges that, in view of the importance of some programs, adequate observing time be granted on these telescopes in the early 1980's, while waiting for the construction of new astrometric telescopes. Such projects might include parallaxes of faint stars, notably at the lower end of the H-R diagram (stellar evolution), accurate measurements of the internal motions in clusters (dynamical models), parallaxes of some selected stars, and the measurement of double stars by visual and interferometric methods (stellar masses).

As mentioned above, photoelectric detection methods currently under development will be considerably more precise than current photographic methods. Whereas such methods could possibly reach accuracies of the order of 0.1 mas, even the best existing telescopes are probably limited to 1 mas. It is therefore important to incorporate as well as possible the ideal design characteristics listed above in any new telescope design. In addition, the relative complexity of photoelectric systems will probably require a constant temperature and gravity environment, i.e., a fixed focus, in order to reach their ultimate accuracy. Two promising designs, both featuring only one moving plane mirror, have been proposed. The Working Group believes that further design studies should be conducted leading ultimately to construction of ground-based astrometric telescopes in both the northern and southern hemispheres in the mid-1980's. Technical studies indicate that a 1-m class telescope allowing 0.1 mas is technically feasible. The effects of the Earth's atmosphere allow an accuracy of 1 mas per night, leading to 0.1 mas for parallax determinations. Although the number of stars measured with such telescopes is very much smaller than the current output of the 1.55-m USNO astrometric reflector, reliable parallaxes at 1 kpc (10 percent error) will be obtainable.

The ability to achieve a parallax accuracy of 1 or 0.1 mas depends also on the determination of the correction from relative to absolute parallax for each field to even better accuracy. Since the typical correction to absolute parallax for faint stars is approximately 1 mas, we expect the error of the absolute parallax to be dominated by the error of the correction and not the error of the measured relative parallax. Two approaches may be used to determine the correction:

1. The method of spectrophotometric parallaxes; and
2. Deconvolution of the measured distribution of the relative parallaxes of the reference stars and their errors to obtain the correction.

In some cases, the measured relative parallax of a nearby quasar or star known to be very distant will provide the correction, but these cases will be rare, and the accuracy will be limited to the accuracy of the quasar's measured relative parallax. The on-line measuring devices proposed to date are limited to approximately 10 reference stars, which eliminates the deconvolution method as a possibility since several hundred reference star parallaxes are required to define the distribution accurately. On the other hand, if we assume a correction to absolute of 1 mas for 16th magnitude, and 10 reference stars, we find an error of 0.1 mas for the correction. It seems unlikely that deconvolution of the parallax distribution in the photographic cases, where many reference stars are available, will provide a more accurate correction.

It seems, therefore, that the usefulness of relative parallaxes will be fundamentally limited to accuracies of somewhat less than 0.1 mas. Our conclusion is that parallaxes with accuracies of 10 percent or better will be limited to distances of equal to or less than 1 kpc, except for cases in which the star is measured relative to a bright quasar or objects known to be distant (such as stars in the Magellanic Clouds).

Finally, it should be noted that such specialized telescopes, which are able to reach 0.1-mas relative positional accuracies, will allow a vast improvement in our knowledge of the nearby stars. A survey of the known stars within a given volume (say, the 500 or so stars within 12 parsecs of the Sun) with a precision more than 10 times that available during the 1970's will reveal the following:

1. The vast majority of the currently undetected stellar companions (the exceptions should be detectable by radial-velocity studies);
2. The frequency of occurrence and mass distribution of substellar ultraplanetary masses (10-60 Jupiter masses); and
3. Virtually every planetary system with a planet exceeding 0.2 Jupiter mass and an orbital period in excess of 10 years (in some instances planets as small as 0.02 Jupiter mass could be detected).

Besides the detection of new objects, such telescopes would measure the characteristics of the visible objects to unprecedented precision. For example, the errors in the luminosity arising from errors in the measured parallax would be reduced to 0.5 percent or less, and in some cases to as little as 0.05 percent. The errors in stellar masses will be similarly affected. Because the main source of error in these calculations has always been the error in the parallax, a number of masses with an error of only a fraction of one percent will become available to theorists during the first few years of operation. Thus, these expected new developments in telescope construction will have a major impact on many areas of astrophysics.

C. Fundamental Astrometry Instrumentation

The global large-angle measurement characteristics of fundamental astrometry lead to unique instrumentation requirements. The fundamental reference system is based exclusively on absolute transit-circle meridian observations. These observations result in catalogs such as FK4 and the forthcoming FK5, which are the closest approximations to an inertial reference system available. At epoch, the mean errors of such absolute catalogs amount to perhaps a few hundredths of an arcsecond. Errors then accumulate because of errors in the proper motions. These errors are being reduced, and the reduction process can be accelerated, by incorporating new technology into the classical observing systems, because of the following:

1. New large-angle measuring techniques. Highly precise and accurate divided circles are probably near their ultimate limit around 0.05 arcsec for a single, multistation reading. The use of inertial and/or ring-laser gyro methods appears promising and should be investigated.
2. Improved focal-plane mensuration and integration methods. Photoelectric devices in general have made valuable contributions, and solid-state photosensitive devices, as discussed earlier, offer great future promise.
3. Atmospheric investigations.

D. Fundamental Interferometry Instrumentation

The high-resolution characteristic of interferometers has been successfully used at various wavelengths, viz., measurements of source structure at radio frequencies; observations of circumstellar dust shells in the infrared; and of course the early, and recently refined, determinations of stellar diameters at

visible wavelengths. The adaptation of this resolution to
astrometry, and in particular to large-angle astrometry, has
progressed unevenly in these various spectral regions. In
general, the longer the wavelength, the further the development
has been carried. This is to a great extent (but not exclusively) due to the increasing complexity of the instrumental
techniques as one goes to shorter wavelengths.

1. Radio Astrometry

Radio astrometry has already produced excellent results. The
methods currently used can measure absolute declinations and
absolute right ascension differences. Observations of minor
planets and/or pulsars may allow a determination of the equinox.
Depending on baseline orientations, geometrical correlations
can cause the precision to vary with position. Variable source
structure appears to be nearly the rule at the milliarcsecond
level.
    The questions to be addressed in the 1980's include the
following:

    • The continuing identification of those sources most
amenable to astrometry;
    • Improvements to the corrections currently applied for
the various atmospheric effects, which currently limit accuracies to a few hundredths of an arcsecond;
    • The interelating of the optical and radio reference
frames on a global basis—in this way the assumed inertial
characteristics of the quasars can be most profitably used; and
    • The commitment of radio facilities, in a predictable,
continuing fashion, to longer-term astrometric programs,
including those in the southern hemisphere.

2. Infrared and Optical Astrometry

Infrared interferometry can utilize the heterodyne techniques
used in the radio regime. The fringe-phase technique used in
radio astrometry is applicable here, and the observed quantities
are generally the same as in radio. Infrared also has the
advantages of "medium-scale" wavelength, and, perhaps most
importantly, it is relatively indifferent to the composition of
the atmosphere, i.e., to the partial pressures of dry air and
water vapor. In addition, many fundamental stars can be
observed directly. Despite these apparent advantages, no
large-angle infrared astrometry has been accomplished to date.
    Neither have large-angle astrometric observations been made
by an optical interferometer. Progress regarding instrumentation is, nevertheless, being made. The problems of baseline
stability, common to all interferometers, is naturally more
difficult at these wavelengths. In addition, the composition

of the atmosphere must be taken into account, even if two-color interferometers are used. It is clear, however, that both instrumental and atmospheric problems can be reduced; the problems are understood, and solutions have been proposed.

In view of the tremendous importance of these techniques in the determination of fundamental stellar positions and motions, down to the 10-mas and possibly 1-mas level, their use in determining Earth rotation parameters and the additional small-angle capabilities (stellar structure and diameters), the development of multiple-element interferometers in the 1980's should be strongly supported.

E. Atmospheric Investigation

All types of ground-based astronomical observations are affected by atmospheric effects to a greater or lesser extent. As far as large-angle astrometry is concerned, unmodeled refraction and phase effects can cause systematic errors in both direct-imaging and interferometric methods. These errors range from 0.005 to 0.1 arcsec, generally averaging to a few hundredths of a second of arc. In short, it is "easy" to account for 99.8 percent of these effects, but when one talks of centiarcseconds or milliarcseconds, the remaining 0.2 percent becomes important. Complementary improvements can be made as follows in the 1980's:

1. Atmospheric probing, with particular emphasis on water-vapor profiling. Several methods--acoustic, radiometric, and lidar--are possible. The last is probably the most promising for real-time atmospheric probing.

2. An allowance for anomalous refraction (i.e., isopycnic tilt) by either two-color dispersion measurements or by atmospheric probing as in 1 above. The two-color method is intended to measure the total refraction regardless of origin, but the proper index of refraction must be known, which necessitates a knowledge of water-vapor content. In addition, nonzenith two-color observations must be corrected for the effects of a spherical atmosphere, and at large zenith distances (in common with any approach) for atmospheric structure.

Note that a new theory of refraction is not required, but rather the use of more observed information in the reduction of observations. Obviously, such atmospheric studies will also aid in the corrections to spectroscopic and photometric observations.

F. Measuring Engines

While instrumental developments during the 1980's will undoubtedly see the mounting of photoelectric or solid-state measuring devices directly in the focal planes of telescopes, such devices will probably be restricted to small angular fields in the sky. The observation of larger fields will still require the use of conventional photography and the subsequent measurement of photographic plates on some form of measuring engine. Future astrometric programs will also require the remeasurement of some of the extensive plate collections already in existence. Accordingly, there is a continuing need for the development of measuring engines of ever greater speed, accuracy, degree of automation, and capacity for on-line storage and analysis of data.

At present, there are only three automatic measuring engines in the United States that are dedicated to astrometry--two at the USNO and one at the Lick Observatory. A new machine now undergoing commissioning tests at Yale Observatory will be partially used for astrometry. All four of these machines are fully committed to research programs that represent only part of the total activity going on in the field of astrometry. Microdensitometer-type measuring machines in other institutions, if carefully adjusted and maintained, could in principle also be used for astrometry. However, the need for dedicated astrometric machines and for measurements of higher precision in the 1980's leads the Working Group to conclude that three additional measuring machines, incorporating the latest state-of-the-art instrumental design and techniques, should be funded and built.

At the same time, the creative use and upgrading of existing conventional machines should be supported and funded. As examples, we note the experimentation with solid-state array detectors and the possible application of high-speed flying-spot scanners to precision measurement.

Finally, the success of the Control Data Corporation automatic plate scanner in blinking plates for a proper motion survey of very faint, high-proper-motion stars should be specially noted. Every effort should be made to extend the capabilities of this unique machine to other survey-type projects, such as the detection of variable stars and very blue or red stars.

G. Space Astrometry

Two major sources of error in astrometry are atmospheric and flexure effects. Astrometric observations made from space eliminate the former and greatly reduce the latter. The first astrometric efforts in space are scheduled to be made during

the 1980's and consist of NASA's Space Telescope (ST) and the
European Space Agency's High Precision Parallax Collecting
Satellite (HIPPARCOS).

The primary astrometric instrument of ST is the Fine
Guidance System (FGS), which is designed to permit relative
astrometric measurements of objects in the magnitude range 10
to 17, over a 60 square minute-of-arc field to an expected
accuracy of ±2 mas. Additional astrometry will be possible
with one or more of the imaging cameras available for use on
ST. These relative astrometric capabilities are anticipated to
result in parallaxes at or slightly below the millisecond of
arc level of accuracy. The lack of a large-angle measuring
ability severely restricts ST's application to fundamental
astrometry; however, it may provide some information useful for
the adjustment of ground-based fundamental astrometric observations. The high resolution of ST will bear on the problem of
close binaries (about 30 mas) using a variety of instrumentation that includes the FGS and imaging cameras. A related
effort involves the search for unseen planetary-mass companions.
A final area of astrometric interest is the occultation of
stars by solar-system objects.

HIPPARCOS, if successfully built and launched, will measure
a large fixed angle (about 70°). Thus, large arcs between
suitable pairs of objects can be measured. These arcs then
will form the basis for a global solution for positions,
motions, and absolute parallaxes for some 100,000 stars with
blue magnitudes less than 11. The positions and motions are
described as defining a "rigid" system with an unknown orientation and residual rotation. Although, therefore, not a fundamental system, if the predicted accuracy of internal position
and proper motions is reached (±2 mas and ±2 mas/year) the
system can be used in an interactive way with ground-based
fundamental astrometric observations in order to improve both
the ground and space systems. HIPPARCOS is primarily an absolute parallax-measuring satellite with relative positions and
motions derived as necessary "by-products." HIPPARCOS, it is
expected, will also discover thousands of close binaries.

Both ST and HIPPARCOS represent valuable first astrometric
steps into space. The former is primarily an astrophysical
instrument with limited astrometric potential, while the latter
would be truly the first dedicated astrometric satellite. The
experience gained from these first efforts will lead to further
improvements and capabilities. ST may be characterized as
"small-angle relative," HIPPARCOS as "large-angle relative."
An obvious extension involves an "absolute" capability coupled
with the dynamic range necessary for directly relating the
radio and optical reference frames on a global basis.

Shuttle-based systems and free-flying satellites will both
have tremendous potential in all the various areas of astrometry. The Working Group favors the continued study of long-

focus astrometric telescopes with sophisticated focal-plane devices for Shuttle-based systems and development of such a telescope for deployment during the 1980's. Such deployment should be a forerunner to a dedicated free-flying astrometric satellite with the capability of measuring arbitrarily large angles on request (requiring the development of large-angle measuring devices), coupled with sophisticated focal-plane measurements. Such an astrometric space telescope will revolutionize both fundamental and relative astrometry in a comprehensive and complementary fashion. It is not clear at this time whether such an ultimate astrometric telescope would take the form of an interferometer, a single-focus instrument, or a combination of these. But, in any case, it is already clear that accuracies in the fundamental system of 1 mas, or perhaps considerably better, will become a reality in the 1990's if studies and prototypes are funded in the 1980's. Preliminary studies indicate that 1-microarcsec accuracies in relative motion might not be out of the question, thus opening up the Magellanic Clouds to study and allowing the detection of Earthlike planets in the solar neighborhood.

7

Search for
Extraterrestrial Intelligence

I.   SUMMARY AND CONCLUSIONS

The Working Group on the Search for Extraterrestrial Intelligence (SETI) has reviewed the arguments for and against a
SETI program during the 1980's, taking the substantial body of
literature on SETI as background for this task. We have
attempted to remain mindful of the wide range of controversy
concerning the prospects for the detection of nonterrestrial
intelligent life. While we recognize that there is as yet no
direct scientific evidence for its existence, we believe that
the detection of extraterrestrial intelligence would constitute
one of the most profound discoveries of this or any age.
Our conclusions are as follows:

1.   The Working Group believes unanimously that searches
for extraterrestrial intelligence are both justifiable and
timely for the 1980's.
2.   SETI activity should be undertaken on a long-term
basis, involving the participation of the general scientific
community, rather than as a single project of fixed duration.
Most arguments of which we are aware suggest that the chances
for ETI detection in any few-year time period are small;
negative results are to be expected in early phases of the
program.
3. Only modest funding for SETI is warranted at present,
because of the presumably small chances for early success of
SETI and because of the competition for funding by other
high-quality programs. Moreover, a portion of those funds that
are committed to a SETI program should be reserved specifically

to support original or unconventional approaches to overall
SETI program goals.

    4. One worthy program for the 1980's is the study of
signals in the microwave region of the spectrum in support of a
SETI effort rather than for possible benefits to any area of
astrophysics. However, innovative projects in other frequency
ranges should be considered and evaluated on their merits.

    5. Concurrent with a SETI initiative, a program for the
detection of extrasolar planetary systems should be vigorously
pursued. Such a program would not only be relevant to SETI but
would also address a range of important questions in Galactic
astronomy and astrophysics. Astrometric techniques appear to
offer a promising approach to the detection of extrasolar
planets.

II. BRIEF REVIEW OF WORKING-GROUP DISCUSSIONS

A. The Concept of Special SETI Searches

Astronomers are continuously making observations of ever-
increasing refinement, which, in principle, could detect some
evidence of alien life. The real issues are whether <u>specific</u>
searches for extraterrestrial intelligence are appropriate for
U.S. astronomy in the coming decade and, if so, by what means
and on what scale. Related issues include possible spinoffs to
radio astronomy and other areas of science and technology,
together with such questions as the abundance and characteristics of planets around other stars.

    The Working Group is unanimous in the belief that the time
has come for sensitive and long-continued SETI programs to be
undertaken. Technology is now available, at a reasonable cost,
to make meaningful searches in a much more effective way than
would be possible through the use of conventional astronomical
instruments and programs, with reliance on chance alone.

    It is hard to imagine a more exciting discovery, or one
that would have greater impact on human perceptions, than the
unambiguous detection of extraterrestrial intelligence. On the
other hand, the complete failure to find any evidence of alien
life, after long and careful searches of many kinds, would have
important implications and probably consequences.

B. General Context of the SETI Activity

Certain arguments combine to suggest that the likelihood of ETI
detection in any few-year time period is very small. These
include considerations of interstellar distances, the vastness
of the multidimensional SETI-parameter "phase space" (see
Section D), the speed with which a single energetic species

could populate an entire galaxy (and if this has been done in our own Galaxy, where are they?), and the uncertainty of assumptions that our own technology is similar to that of possible other civilizations. Because of this presumed small chance of early success, together with the severe competition for funds from expenditures with demonstrable short-term payoffs, SETI should be modestly funded at this time. Moreover, it should not be thought of as a "project," but rather as a long-term activity that will surely evolve and perhaps even change radically in its search techniques, and that may involve a number of specific programs directly and indirectly related to SETI (see Section E).

In this context, we note that NASA's "Life in the Universe" program includes SETI activity. We support NASA's current SETI Science Working Group approach designed to open a microwave (indeed any reasonable) SETI program to continuing competition and initiative from the full scientific community, with peer review; to encourage imaginative and low-cost approaches to the development of SETI hardware and software; and to consider carefully the tradeoffs of scope and scale versus cost and performance.

C. Possible SETI Approaches for the 1980's

We are agreed that the microwave approach offers great promise in the coming decade. However, four general classes of potential SETI opportunities were discussed:

1. Physical Presence

This category might include, for example, the search for artifacts, "DNA capsules," and UFO's. These are controversial subjects, which for adequate treatment would involve fields other than astronomy and would require a far more time-consuming study than the Working Group was able to undertake. Accordingly, we did not consider such topics to lie within the scope of our charge, apart from the possibility (in principle detectable with astronomical techniques) that one or more minor bodies of our solar system may have been altered or may even be artificial.

2. X-Ray, Gamma-Ray, or Optical Emissions

Advanced technologies might well rely heavily on high frequencies for signaling and communication because of the high information-rate possibilities and the relatively small physical size of systems able to generate tight beams. Unfortunately, many natural sources are quite bright over much of this frequency range; also the tightness of possible beamed transmissions makes the chance of accidental eavesdropping very small. Although an open mind should be kept for good ideas, at least for the present this may be left primarily as an area for accidental discovery.

3. Infrared and Submillimeter

This category includes at least three quite different types of possible indicators of extraterrestrial intelligence:

(a) The infrared and submillimeter spectral region appears to be favorable for interstellar communication on signal-to-noise grounds but along with the higher frequencies may suffer from difficulties in our eavesdropping on beamed transmissions.
(b) Laser power beams in the multigigawatt range have already been contemplated by our youthful technology; advanced civilizations might make widespread use of those in far more powerful forms, with spillover or scatter detectable at some distance.
(c) Waste radiation from a civilization that utilizes a large fraction of the energy emitted by its central star would be detectable as a mixture of lower-temperature flux from the artifacts and high-temperature flux from the star.

During the coming decade our technology in this spectral domain will not be sufficiently ripe to conduct a major search for any of the above. Initial infrared surveys able to reach sources faint enough to be interesting will be made during the 1980's; in the following decade it may be appropriate to consider the question of deeper, SETI-motivated infrared and submillimeter searches.

4. Microwave

Any major SETI program in the 1980's should concentrate in the centimeter and perhaps millimeter region of the spectrum, for the following reasons:

(a) Existing antennas can nicely handle this kind of search.
(b) Our microwave technology is now advanced enough for near-optimum receivers and correlators to cover substantial amounts of the SETI "phase space."
(c) Under plausible scenarios, for a given amount of transmitted power the received signal-to-noise ratio should be most favorable in the 1-100-GHz region.
(d) At least by our present understanding of technology, some types of powerful "broadcasts" on which we might hope to eavesdrop (such as relatively broadbeam radars or scatter from power transmissions) are likely to lie in this general part of the spectrum.

D. A Possible Microwave SETI Program

Previous SETI attempts have explored only very small slices of the potential SETI parameters of frequency, bandwidth, direction, intensity, pulse duration, and search time. Six or more orders of magnitude improvement over previous searches can now be achieved, for example, by an approach being developed at the NASA Ames Research Center in cooperation with the Jet Propulsion Laboratory (JPL), wherein near-optimum low-noise amplifiers would feed 10^6-10^7-channel spectrum analyzers with resolution from 1 Hz to 1 kHz. The system would include novel pattern-recognition algorithms of a type far more effective than those used in previous SETI searches. The full system as currently proposed by Ames-JPL for the 1980's would be powerful and efficient, although we believe it is likely that a cost-effective program of this type could be carried out at a somewhat lower level or rate of expenditure.

This activity could make modest but helpful contributions to radio astronomy. Some useful equipment would be produced, such as broadband low-noise front ends, powerful spectrum analyzers, and data-handling units. The discovery of new spectral lines, pulsars, and other astronomical sources can confidently be expected, and there is the further possibility of discovering totally unexpected natural phenomena producing extremely narrow spectral lines. Other branches of science and technology, such as seismic surveying, may also profit from some of the technology developed to handle and analyze the SETI data. However, we do not believe that a large fraction of the cost of such a program could be justified by its spinoff applications to radio astronomy and other fields.

E. Other Studies Relevant to SETI

Other studies relevant to SETI include those on biogenetic molecules here and elsewhere in the Universe, as well as on evolution of life and its effects on planetary atmospheres and surfaces. Beyond giving general encouragement, we do not feel competent to offer advice in these areas. Two other areas do, however, fall within our astronomical purview.

The first, mentioned earlier, is the question of whether astronomical searches for extraterrestrial intelligence should include the solar system. The Universe is, after all, very old. According to some scenarios, ETI may be ubiquitous. During the past several billion years, our solar system may have been visited before--perhaps often. Interstellar travel, at least as now generally conceived, would appear to demand large quantitites of matter if only for reaction mass, together with heavy reliance on utilization of matter and energy at the destination. Some reasons have been given to suspect that minor bodies of the solar system could be attractive sources of materials or could even be altered to serve as "stations" for travelers. However unlikely these cases may appear, it may still be worthwhile to extend the studies of minor bodies in the solar system and to look closely at any anomalous ones.

A second program, and one certain to yield scientifically useful data, is the search for planets around other stars. Although most astronomers believe that planetary systems must be common, we have as yet no conclusive evidence for even a single one apart from our own. It is important to begin a major program during this decade to study the nearby stars, binary as well as single, from this point of view. Several approaches to extrasolar planetary detection, including radial-velocity studies, should be supported; however, we believe that, over the next several decades, astrometry is the approach most likely to develop the potential for the detection of planets much smaller than Jupiter.

New astrometric techniques using old telescopes already permit work on selected nearby stars at the several-milliarc-second level. One or more new telescopes of modest size and relatively low cost, specifically designed and located to facilitate this work, should give an order of magnitude further improvement in accuracy. Support for these is needed, not only for the planet-detection problem but also for the detailed information such astrometric measurements will yield on the frequency and nature of binary systems, on precise masses of stars, and on very accurate parallaxes and proper motions, which will markedly improve our calibration of distance scales in the Universe. Several further orders-of-magnitude improvement in astrometric accuracy will come with the use of appropriately designed space telescopes, making it possible to detect Earth-sized planets out to 10 light-years. Study and development of such systems should be pursued.